'Water voles are adorable little beasts. They are also tough, randy and stroppy, as Tom Moorhouse makes clear in this wry, amusing account of the often bloody, painful and frustrating business of conservation fieldwork. Accounts of the decline and degradation of wildlife and the natural world are generally off-puttingly gloomy, but Tom Moorhouse handles his subject with a light, self-deprecating touch. "I hold stubbornly to optimism," he declares, and his "Elegy for a River" demands that we do the same.'

CHRISTOPHER SOMERVILLE, *THE TIMES* WALKING CORRESPONDENT, AUTHOR OF *THE JANUARY MAN*

'Oh my ears and whiskers. I loved this . . . Self-deprecating humour combines with a paean to the wonders of creation, hard facts and hope for an imperilled species.'
SAGA MAGAZINE

'Tom Moorhouse has written a book about ecological loss that is also somehow laugh-out-loud funny – passionate, warm and full of fascinating insights into the eccentric world of the field naturalist.'
ISABELLA TREE, AUTHOR OF *WILDING*

'A fascinating story of ecology and fieldwork that is both funny and furious. Moorhouse has written an elegy not just for the rivers he loves, but also for life on earth.'
HUGH WARWICK, AUTHOR OF *A PRICKLY AFFAIR*

'Beautiful and important. Tom's book is extraordinary in its gentle curiosity and sympathy for his subjects. In *Elegy for a River* he takes us back to our childhoods. He then holds our confused moral compass up to a microscope to make us realise that only a return to that place can save us. I love this book.'
SIR TIM SMIT KBE, EXECUTIVE VICE CHAIRMAN AND CO-FOUNDER OF THE EDEN PROJECT

'What a book. It has everything I love. It is lively, it is tender, it is fascinating, it starts small and very particular, and then – my God – by the end you are doing the Hallelujah chorus. It feels such an important book and I hope that everyone reads it. It seems to me to deliver on the greatest thing a book can achieve – when, through reading, you feel changed and inspired to act.'
RACHEL JOYCE, AUTHOR OF *MISS BENSON'S BEETLE*, AND *THE UNLIKELY PILGRIMAGE OF HAROLD FRY*

'It flows from the heart, eddies with fascinating information, and runs cool and clear with concern about the state of our rivers. They now have their champion.'
JOHN LEWIS-STEMPEL, AUTHOR OF *MEADOWLAND*

'Small is beautiful. That goes for conservation, too. Tom Moorhouse doesn't chase tigers and elephants; he gets bitten by water voles. He doesn't big up the case for saving the Amazon; he pleads for the tiny streams and forgotten pools where his voles swim and scamper. *Elegy for a River* is an unabashed love story about a soggy, decade-long adventure into the heart of the watery English countryside, in search of his wild, wonderful and sharp-toothed obsession. A joy.'
FRED PEARCE, AUTHOR OF *WHEN THE RIVERS RUN DRY*

ELEGY FOR A RIVER

Whiskers, Claws and
Conservation's Last, Wild Hope

TOM MOORHOUSE

PENGUIN BOOKS

TRANSWORLD PUBLISHERS
Penguin Random House, One Embassy Gardens,
8 Viaduct Gardens, London SW11 7BW
www.penguin.co.uk

Transworld is part of the Penguin Random House group of companies
whose addresses can be found at global.penguinrandomhouse.com

First published in Great Britain in 2021 by Doubleday
an imprint of Transworld Publishers
Penguin paperback edition published 2022

A CIP catalogue record for this book
is available from the British Library.

ISBN
9781529176728

Typeset in ITC Berkeley Oldstyle Pro by Jouve (UK), Milton Keynes.

Printed and bound in Great Britain by Clays Ltd, Elcograf S.p.A.

The authorized representative in the EEA is Penguin Random House Ireland,
Morrison Chambers, 32 Nassau Street, Dublin D02 YH68.

Penguin Random House is committed to a sustainable
future for our business, our readers and our planet. This book
is made from Forest Stewardship Council® certified paper.

In memory of Rob Strachan – the original,
and finest, Volemeister

CONTENTS

1

A Wild Hope

You may say I'm a dreamer,
but I'm not the only one.

JOHN LENNON,
'Imagine'

I SPENT YEARS TRYING TO AVOID THAT 'SO, WHAT DO YOU DO?' conversation. It always turned a bit awkward.

'I go out and study water voles in the wild,' I'd say, 'trying to conserve them.'

Which usually prompted confusion. 'What, waterfalls?'

'No,' I'd smile. 'Water voles. Cute, endangered, rat-type animals.'

'Oh.' A moment's thought. 'Why?'

I got that a lot. And there are few explanations that don't sound evangelical.

The other common response was, 'Oh, wow, that's great! It must be lovely being outside all day rather than stuck in an office. I'd give anything to do that. You're so lucky!'

And I know I should have just agreed and moved on, but I could never manage it. I'd settle for something like, 'Yeah, sure, but there are drawbacks. I mean, it does rain a lot.'

And in return I got a look that said I'm an ingrate.

But what I really wanted to say was, yes, of course you're right. It's wonderful. I get to spend all day with wildlife, doing what I always dreamed of. How amazing is that? But, look, it also means I've dedicated my life to the service of Fieldwork. And Fieldwork is a goddess with a mean streak and a sense of humour. In lieu of blood sacrifices – of which she receives plenty, thanks to her unparalleled potential for causing minor personal injury – she'll also happily accept a researcher, head bowed to clenched fists, muttering a prayer of 'Why am I doing this?'

At some point in every fledgling ecologist's career they discover the nature of the deal they've struck – by which time it's far too late. In my case, I'm only proud that it took her so long to nail me. But then Fieldwork has patience. And she starts gently. She lures you, first, with beauty and intrigue. She snagged me aged twenty, when I spent my summer as part of a team surveying river vegetation. This meant getting dressed up in a fetching yellow-and-black drysuit and wading down predetermined lengths of the river Swale, in the north-east of England. The upland sections were distant, bleak and beautiful. The river there was all peat-brown rapids, and rocks with clinging trails of river mosses and enclaves of liverworts. We needed guide books and microscopes to identify some of the tinier, more obscure plants. It was hard work. But then 'work' also meant jumping into upland pools, eating lunch by water-falls, finding the picked-clean remains of an otter's dinner strewn across a flat boulder, or catching a cobalt flash as a kingfisher zipped by.

In the lowlands the river became deep, winding and serene. Here plants clustered in dense stands and were more easily identified. We ticked off species from a list on a clipboard as we went. And in the lower reaches we could float on our backs, buoyed by our inflatable chests.* Borne through dark waters on a stately current, we gazed up at a towering canopy of trees. We drifted in peace. Right up until the rubber neck-seal on my drysuit ruptured, midstream. Air hissed out, and

---

* I should probably explain. The river squeezed air up out of the drysuit legs, handily turning the arms and chest into a giant flotation device. Bobbing along, we each looked like a cross between a duck, a bee and half a Michelin Man. It's normally safe enough to float in drysuits, in which you can avoid the reverse process that often befalls those who opt for chest waders. More than one chest-wadered researcher has been swept downstream dangling from a pair of inflatable feet.

river poured in. My boots filled with water and I started to sink. Far from the bank and out of my depth, I did what anyone would: I panicked. I flailed, paddled, shouted, struggled, whimpered and scrambled my way to the shore. I crawled out, peeled off the drysuit and spent the remaining day in soaked-through clothes. Not fun, but one of those things. Good attempt, Fieldwork.

On the intrigue side, Fieldwork can be relied on to provide the spectacularly weird. Downstream from Richmond we once spotted a polystyrene takeaway tray on which somebody had placed – for no adequately explicable reason – some poo.* They had adorned this with a small Union flag, fixed at a jaunty angle, and set the whole thing sailing off down the river. We watched it until it was gone from view. On a similar theme we discovered an encampment of total-immersion baptists on the river Wiske. The baptists spent their days in the river, dunking one another to wash away their sins. They were bemused when three drysuited researchers, laden with gear,

---

* The exact species of faeces remains a mystery. I mentally narrowed it down to human or dog but wasn't about to investigate further. And I know we probably should have caught it and cleared it up . . . but yeah, no.

clambered up the riverbank and lurched apologetically off through their camp. We were bemused by their choice of location. The Wiske is a short river with multiple sewage-treatment works. The water was . . . nutritious. Really, it was just bursting with well-fertilized plants. Working there meant strict hygiene protocols, including rubber gloves, laboratory soap and cans of tapwater to wash with before eating or getting in the car – all to minimize the chance of ingesting something disastrous. The baptists were nearly naked and fully submerged. If their aim was to get closer to God, they risked being a lot more successful than they had probably intended.

The winter after that project was spent sitting in a freezing bird hide, recording blue tits, great tits and coal tits as they chased one another off bird-feeders. (All while enterprising squirrels hung upside down from tree branches and chewed the feeders' ropes, so they could ruin the experiment and get at the peanuts when the feeders hit the deck.) It was so cold that our electric heater cracked the window. Bitter, yes, repetitive, yes, annoying on the squirrel front, definitely. But still quite lovely. And a year later I participated, sleep-deprived and sunburnt, in dawn-to-dusk surveys of summer pollinator abundance. That's what they call 'type-two fun', great in retrospect. And afterwards I spent months literally over my head in razor-sharp fenland vegetation, surveying plants at Wicken Fen. The problem with fenland grasses is they flatten to cover the ground ahead of you as you walk. Which is why I confidently strode on to something that turned out to be a six-foot-deep water-filled trench. My friend thought it was hilarious. I didn't. But it was just another soaking. The fenlands were beautiful, the weather was warm and I still loved Fieldwork. So when in June 1999 I was offered the opportunity for a doctoral study of water voles in Oxford, I barely even thought before saying yes. I'd never caught a wild animal before, and was apprehensive about that. But I was confident I could cope. The voles would be a challenge, but I already had lots of field experience, right?

Yeah. Fieldwork broke me in the first week. It was August and rainy. Fittingly, given that I was studying the very same 'Ratty' made famous by Kenneth Grahame in *The Wind in the Willows*,* I had to work from a rowing boat, albeit on a canal,

---

* Kenneth Grahame has dogged my professional life. Or, more accurately, voled it. There remains no way to describe my work without referencing Ratty: yes, he's a water rat, no, that's not a rat, yes, they look superficially similar because they're both rodents, but

not a river. I'm not a natural rower but could get the thing
going mostly the right way. I had just finished an intensive
week's worth of vole-handling training. And that, I thought,
meant I was pretty much sorted.

I wasn't. And I have since developed advice about catching
wildlife from a rowing boat: don't. Trying to weigh and tag
a struggling animal from a platform that's threatening to
untether itself and drift off into canal traffic falls squarely into
the category of human endeavour known as 'less fun than it
could be'. I spent a lot of time variously clutching at bankside
plants because someone had hurtled by with their boat at full
throttle, or trying to get rid of tipsy champagne drinkers who
pulled alongside in rented narrowboats and demanded to
know what I was up to and whether I was 'going to eat a water
vole sandwich'.* As a consequence of my still-rudimentary
vole-handling abilities my fingers were covered in bites and
therefore plasters, most of which got bits of straw and mud

---

although rats can live by water, and do, which is annoying of them,
water rats are still not rats but water voles, which are a different
animal and from an entirely different rodent family, and which have,
incidentally, been here for thousands of years longer than rats. Clear?
* 'No.'

stuck in them. And pretty much every vole I picked up urinated down my fleece. But the rain washed most of that off, so I guess that's something. Added to all this I forgot to confirm my booking at the local B&B and so had to commute 80 miles from Oxford for an 8 a.m. start every day. The week was very, very long. Somehow, though, I made it to Friday – which was good, because it meant I could collect my traps, finish around midday, pack up and go home for a rest.

None of which happened. I spent Friday afternoon kneeling in a muddy puddle on the boat's plank floor because I'd fumbled my last vole of the day.* It chewed a hole in the handling net, slipped out and leapt for freedom. It hit the planks and bolted for the nearest cover, which happened to be the giant tangle of cage traps I had been methodically stacking beneath the boat's foredeck. A wriggle and a twist, a tail slipping from view, and my day became an exercise in rodent retrieval. An hour later I wearily lifted the final trap free and

---

* 'Believe me, my young friend, there is nothing – absolutely nothing – half so much worth doing as simply messing about in boats.' Kenneth Grahame actually wrote that. To which I can only respond: 'Really? Try it.' Although, to be fair, a lot depends on what the messing about entails.

chucked it on to the mound on the canal bank. Revealed in the empty space, facing me and spread-eagled like a rock climber high in the boat's prow, was my water vole. Its expression, above braced legs and fluffy, pale tummy, was defiant. It squeaked and bared orange teeth. I grabbed for it. It dropped, dodged, then bit deep into my leather glove. It held on just long enough for me to manoeuvre it over the side and release it to the dark safety of the canal.

I took a breath. Then I got on with repacking all the traps.

On my way back to the dock, one of the boat's rowlocks snapped. And it started to rain again, hard. Through torrents I punted my boat back up the canal, using an oar as a make-shift pole. On the last curve I passed a Chinese tourist. He was standing under an umbrella on the towpath, smiling and holding a camera. I was sodden, chilled, late, tired, aggravated, filthy and smelt strongly of vole urine. The tourist took a photograph. I carried on punting to the dock. I tied the boat up, bailed out as much rainwater as I could and slumped in my seat. I lowered my forehead to vole-scented fists. Why am I doing this? I thought.

It's a fair question. And one for which it's difficult to find a simple answer. On one level I guess the bad days have compensations. There is joy in being by water, in working in environments that ripple and flow, in using your body outdoors. That's enough, just, to turn the bad days into necessary sacrifices. They become payment for the knowledge, in my case, that I could catch and handle water voles, that I could get on with my project and hopefully complete my doctorate. But this still makes me sound like an apologist for field trauma.* I think the true reason is deeper, and far more personal. Perhaps it's the only really good answer I have. It's this: because of a wild hope.

Because wildlife is incredibly important to me, and to many, many of us who feel that there should be somewhere and something in the world that humans don't own. Because I think we should always leave space enough for the 'other', the special, the breathtakingly beautiful, the complicated and intricate. As a child I used to sit for hours watching nature documentaries. I drank down all the footage and information about the details of each animal's life. And I longed to witness them for myself. I still do. But across the globe, in every country, the Earth's wild places are being diminished. And with

---

* Stockholm Syndrome, anyone?

the loss of the wildlands comes the extinction of species, each of which is fascinating, charming and irreplaceable. And even though I will never see even a fraction of them, my world is immeasurably better because they still exist. This, more than any other reason, is why I and so many of my colleagues spend our time as conservation researchers. We are trying to work out how to save something we love.

Conservation science is founded on a wild hope. We hope that if we study endangered plants and animals hard enough, we can understand the root causes of their decline in population. We hope we can discover a solution. The challenge is vast and the problems complex, sure, but the concept is simple. Each threatened species is a glittering ecological puzzle box. Somewhere hidden within is the key to deciphering how it works. Research lets us turn the puzzle in our hands, viewing it this way and that. Press here, and here, and perhaps an intricate latch will click open, revealing a secret. Keep going, and just maybe you might find in your grasp the two or three ecological threads that have tangled the species, and the knowledge of how to manipulate them to set it free. Pull a thread. Make an intervention. Save the species, and move on to the next.

This is the deal that lies at the heart of conservation science. We do the research to create the knowledge we need to bring our chosen species back from the edge. We start our projects full of optimism but achingly aware that we could fail. We might spend years in the field or lab and never find a workable solution. That's the risk. All our efforts may end as nothing more than a footnote in the history of something unique that has now passed into memory. But the reward, tantalizingly, is a chance at restitution. Maybe, just maybe, we will be able to return to the field, decades on, and gaze at our study species as it thrives.

Water voles are just one species among the multitude that need our help. They are small, brownish* and innocuous. They don't like being seen. They do like to stamp on their own droppings. They bite. And they are utterly wonderful.

They are found in the northernmost Scottish moorlands where they live in colonies that are mere handfuls of voles, clinging to the banks of streams that are barely more than a

---

* Except in some parts of Scotland and East Anglia, where they can be a striking black. I've never seen one of these melanistic voles, but I really want to.

trickle. These are the black voles, and their fur matches the peat into which they burrow. And in uplands in England and Wales brown-furred voles live in much the same way. Their sparse populations are connected by the intrepid few that hike for kilometres down streams, or even over land, to find new habitats and mates. But as the rivers head down from the high places, their paths soften and broaden. And in these lusher, milder climes the voles' numbers swell in tandem with the plants that now crowd the water's edge. And where the rivers spread further, running even slower and broader through the flatlands, the river voles are joined by throngs more in marshes, fenlands, lakes and ponds. Still water or flowing, it doesn't matter: a thousand busy paws dash along runs beneath tall grasses and succulent herbs. And just a little before the rivers pour themselves out into the sea, side channels split and re-split, forming a marshy web. The flows here stall and reverse, swell and recede, and drift from fresh to salt and back again. Even in these brackish waterscapes water voles carve burrows into salt mud and chew on salty food plants. Everywhere there is water in mainland Britain it is natural to find a water vole living beside it.

They are a totem of our rivers.

And yet ask any class of children if they have seen one and maybe only a hand or two will go up. Had anyone asked me when I was a child, my hand would also have stayed down, but perhaps a quarter of my class would have raised theirs. If my parents' classes had been asked, it would have been half to three-quarters. And when my grandparents were children, every hand would have been raised. Because once, when you walked by water in summer shadows, you would have heard the plop of water voles diving at your approach. Or, if you were quieter, you could have caught a crunch and munch from among the deep grasses. More cautious still and perhaps you would have spotted a V-shaped bow-wave, preceding something small, dog-like and brisk, quickly paddling the water's width. Or sometimes, just sometimes, a water vole would simply sit there on its floating mat of watercress, far beyond reach of the banks, and reflectively gnaw on a stem in full view.

Water voles used to be there. They were ubiquitous but a little reclusive. A sighting was a secret indulgence. Something just for you. A joy.

But now, for most of us, they are a joy denied. This book is an elegy for a river, but the elegy is not for the rocks, or the water or the sediment. Because in those terms rivers are close to immortal. For thousands of years they have sparkled and rippled and glistened in morning fogs. They will continue long after we all are gone. A river is not a flow of water any more than I am a circulation of blood. A river is the rush and bustle of the life that water creates. And few creatures have lived more fiercely or bustled with more determination than water voles. For them those few brief metres of a river's width are a lifeline incised into an uninhabitable desert of moorlands, fields or trees. They and their rivers should always have been inseparable.

A century ago an author turned a common and beloved creature into a character in his book, and water voles became famous, known to millions. By the time I began my research in the late 1990s, they had already nearly vanished. And we knew that if we were ever to be in a position to reverse this loss and restore water voles to their former range, we would require information. We would need to be certain what had gone wrong, to understand with precision the details of the drivers of their decline. Only then could we devise and test solutions that could combat those drivers and begin to put things right. So, yes, that's the reason I gave myself to

Fieldwork. I wanted to help to solve a problem. I wanted my research to return one small part of the natural world to how it should be. It sounds ridiculously worthy, but it wasn't like that. It was selfish.

Because I had a wild hope – that I could bring them back.

2

Fourteen Thousand Years of Belonging

God has cared for these trees, saved them
from drought, disease, avalanches, and a
thousand tempests and floods. But he
cannot save them from fools

JOHN MUIR,
The American Forests

MY FAVOURITE BIT IS ALWAYS WHEN I LET THE VOLE GO.

In my hand is a Pringles tube, within which is my twentieth water vole of the day. I feel her feet patter. The tube tips as her weight shifts. Then her muzzle nudges the leather-gloved palm I'm holding across the tube's open end. She gives me a speculative bite.

'Ow,' I tell her.

She has another go, and this time she means it. Water voles have a well-established strategy known as bite-'em-hard-then-run-like-a-maniac. They have strong jaws for their size – picture a smallish and more even-tempered rat – and I'm grateful for the glove. In a previous year a water vole bit clean through one of my fingernails. The split took six months to grow out, during which I had to superglue the sides together so I could play guitar. I still have a scar. But you can't blame the voles, and this one definitely has a point. There is a limit to how long anyone should be in a Pringles tube. So I set off back through the reeds, following my snail-trail of flattened plants down to the dyke edge where I caught her.

By the time I reach the water, the feel of the fenland has changed. I'm a bare handful of paces from the track where I left my kit but it's still different somehow. This, right here, is her place, not mine. I crouch a moment among the tall grasses, and breathe. Light filters green-gold through leaves. The air is warm, still, earthy, sweet, fresh, moist. Everywhere is the sense of a fen quietly thronging. It's in the rustle and whisper of stems above me, the trilling scatter of birdsong: reed warblers, willow warblers, greenfinches. A buzz – sudden, sharp, receding – as an insect takes flight. And ahead, dark dyke waters chuckle and lap at the peat shore.

She bites me again, bringing me back to business. I gently place the Pringles tube on the ground, close to where the trap was set. Then I back away, and wait. This should be good.

The tube vibrates. A pause. A nose tip eases out. Whiskers follow. They twitch as she scents the air. For long moments my water vole remains utterly still. Then she bolts for the dyke. She's fast, nearly too quick to follow. A dash, a dodge, a leap, a dive, a twist. That distinctive 'plop' as she hits the water. Duckweed bobs, then swirls and closes. And she's vanished – at least from my perspective. In fact, she broke the surface, dived down, kicked up a cloud of sediment from the bed, twisted, changed direction, sped away. Then she barrelled along, tracking the bank towards one of her submerged burrow entrances. She swam up the passage, hauled out and shook herself off. And there she was safe, pattering along her complex of tunnels deep in the earth of the bank where none but the most determined pursuit could follow. This is one of the ways that water voles avoid predators. They are so good at it that Victorian naturalists believed they would hold on to plants at the bottom of the river for hours until the danger had passed.*

I smile, grab the tube and stand up. It doesn't matter that I've handled twenty voles that day, or many thousands in my life – letting her go was a happy experience. It came with a sense of restitution. The world is a little more right than it was. And I will be grateful to get back to my rucksack and gear. To be this close to water is to understand that you are

---

* None of those naturalists, apparently, questioned how long a 300-gram rodent could hope to hold its breath.

out of place. A single clumsy step would prove it beyond doubt. Like this vole, I've spent a lot of time pulling myself, dripping, from fenland dykes. She, however, is designed for it. This stretch of dyke edge, with its thin fringe of water plants, may be stuffed full of predators – foxes, stoats, weasels, herons – but it's also her home. It's part of her territory, a long knife-edge of reeds that makes the difference between survival and death. The evidence is all there. At the water's edge are fresh piles of stems, chopped into lengths. These are her feeding sign. She dropped them while munching on her favourite food plants, chopping their stalks to a neat 45 degrees. And where the bank juts into the water are flattened mounds of her droppings, adorned with a scatter of fresh new ones, for all the world like piles of brownish Tic-Tacs.* She has left these as warnings to other females: my patch – keep out. And leading off, away through the reeds, trails of bare peat have been imprinted on the land by the daily dashing of her feet. Even if I weren't holding a Pringles tube full of water

---

* I tell schoolchildren that water vole droppings are exactly the same size and shape as Tic-Tacs, but that the taste is different. It depends what the vole's been eating, of course, but they're not quite as crunchy and minty. This usually gets the desired response.

vole* I'd know for certain that one lived here. For all the danger, she fits, perfectly. These plants are her food, and this water and earth her protection. She belongs.

'Belonging' is vitally important. Ecologically, belonging is what gives a species the right to survive. To not belong in an environment is to die there. Under natural conditions, any organism that survives and breeds somewhere belongs there – and vice versa. But often, now, we are not dealing with natural conditions. We are dealing with a world shaped by humans. And that makes the concept of 'belonging' layered and murky. For millennia we have hunted animals. Some were useful for their meat or fur. Some were predators, in conflict with our livestock. But in either case, through depleting (and exterminating) populations of born survivors, we have imposed our own definition of whether species belong.

---

* I should probably explain why I had a water vole in a Pringles tube. For years we used nets to keep water voles still while we handled and weighed them. Then somebody realized that Pringles tubes are almost exactly the same diameter as a water vole's tunnel, and that water voles, when stressed, tend to race for their burrow. It turns out that the best way to handle a vole is to point an empty Pringles tube at their head and wait for them to climb in. And then they tend to stay put. Who knew?

And throughout history we have plucked animals and plants from their native habitats and transported them around the world to be deposited into unsuspecting environments they could never have reached without us. Many die immediately. But the effects of those that do not can be catastrophic.* We may wish to undo those effects. And therefore belonging is no longer a measure only of whether a species can survive, but of whether we feel it has the right to.

Which is where things get tricky.

Do rabbits belong in Britain? Most people would probably answer yes. For generations we have eaten them, kept them as pets, watched them hopping around our countryside, written famous books about them, drawn cartoons of them. They are a common, familiar wild British species. But they are not native. They were first brought over by the Romans, and didn't then establish in the wild until about the twelfth century. If

---

* It used to be called the 'tens rule'. Of plants or animals transplanted into a new environment, about 10 per cent will be able to survive. And of those that survive, another 10 per cent will go on to wreak ecological havoc (although this is most likely a gross oversimplification). Much of this book is about my own minor contribution to dealing with the impacts of that latter 10 per cent.

you'd asked King Arthur whether rabbits belonged in Britain (assuming he existed sufficiently to respond), the gist of his answer would have been, 'No, don't be daft.' Or how about brown hares? Or house mice? Or harvest mice? Or fallow deer? Even until recently I'd probably have told you that they are clearly meant to be here. Unlike crazy things like Eurasian brown bears, aurochs, wolves, European elk or lynx. But ecologically I'd have been wrong.

This is where ecology and sentiment collide. In 2017 Lilith the lynx had the temerity to escape from a wildlife park into the Welsh countryside. Her presence was treated with a dismay that verged on hysteria. She was labelled a 'severe' risk to public safety, tracked to the caravan under which she was sheltering, and shot.* Another way of looking at the same event is that lynx are a natural part of the British fauna (they were driven extinct during the early-medieval period) and so

---

* Quick question: how many people have ever been killed, or even seriously injured, by a lynx? It's so close to zero for the difference to be negligible. Far more people have been killed by cows (74 people in 15 years in the UK). Indeed, far more people have been killed by vending machines. And folks actively queue up for those. Nobody, to my knowledge, has had a vending machine shot, just, you know, in case.

Ceredigion council (who took the decision) is responsible for shooting the sole representative of what, for that brief time, was our most endangered native, wild mammalian species. So when it comes to wildlife, 'belonging' means different things to different people. Most of us accept the hares, deer and mice as natural and good, whereas the big, scary elk, bears, wolves, aurochs and lynx are universally felt not to belong. But of course the hares, fallow deer and mice are relatively recent human introductions, whereas the scary things* colonized Britain by natural processes millennia ago, then lived here until we hunted them to extinction. So, tell me, which should we conserve?

We seem to feel that species belong largely if they have been part of our normal experience and are not scary – which argues that nativeness is less important. The counter-argument is that none of us has known a Britain without grey squirrels all over it, and yet, despite their comedy gymnastics, harmlessness and general cuteness, we still might prefer (were it possible) to get rid of the greys and let the red squirrels

---

* By which I mean the things that are extremely unlikely to harm you (compared with, oh, I don't know, bicycles, cars, hair-dryers, domestic boilers, cats, dogs and roller coasters) but of which we still have a bizarre, deep-rooted fear.

return. Why? Because the reds are native, perchance? And don't get me started on our reintroduced populations of European beavers. They are beavers, and so not exactly terrifying, and certainly are native (they may have been at large in Scotland only five hundred years ago). And a generation of Britons is once more growing up in a country in which they live. But current policy is that their populations don't have an automatic right to exist.* They don't really *belong*, you see.

It's a mess. Our decisions are concocted on a species-by-species basis from a bunch of inconsistently applied sentiments surrounding how long something's been around, whether it's familiar, whether it's native, and whether it looks pretty, scabby or frightening. And whether, frankly, we just don't like the cut of its jib. Which is not a solid basis for making choices.

All of this is why I was so grateful to be studying water voles. Because water voles are one of the few wild British mammal species you can point at and say, 'Now there's a creature that really deserves to be here. Yessir, no question. Fits

---

* August 2020 marked the first time a population of British beavers was granted legal permission to remain in the wild – which is wonderful. But the legal status of all the others remains undecided.

right in.' Has it been around for millennia? Yep. Has it been a common part of human experience for generations, including within the lifetimes of living people? Indeed. Is there a world-famous children's book written about them. Yes! Several, in fact.* Do people love them? Clearly, yes, because they are adorable. Do people feel their lives to be threatened by them? Well, in the absence of a legitimate rodent phobia or allergy of some sort, it's tricky to envisage how a water vole could cause you problems, so I'd have to say no, not really.†

---

* You're no doubt (ahem) aware that in addition to the deftly crafted tome you are currently clutching, I have in the past published a series of children's books that follow the adventures of a family of anthropomorphized water voles.* And I'm pleased to tell you that *The River Singers*, and its sequels, have become every bit as famous as *The Wind in the Willows*. In. My. Dreams.

   *Anthropomorphized in the sense that they speak to one another. Not one of them wears a damn waistcoat, though. I was clear on that point.

† They do carry Weil's disease. But a lot of other wild animals also carry Weil's disease. And – I hate to break this to you – a lot of humans you meet also carry some really unpleasant transmissible diseases too. As long as you don't do something dumb, like, I don't know, pursue a career that causes you to permanently have cuts on your hands that regularly come into contact with water vole urine, you should be fine. (And what sort of idiot would do *that*?) If you're close to fresh water, wash your hands before eating. Problem avoided.

In terms of the voles' historic right to be here, you'd need to go back a long way to find a time when they weren't around. At the height of the last glaciation, 22,500 years ago, Britain was essentially a peninsula of northern Europe. The north of what is now our island was permanently snow/ice/glacier-bound. The south was a polar desert: no rainfall, no trees, almost no life to be seen. When the world began to warm again around 16,000 years ago, plants and animals spread from their glacial refugia, recolonizing lands the ice had vacated. At that time Britain was connected to the rest of Europe by a land bridge, and the advancing tide of life swept right on in, water voles included. The oldest British water vole remains come from two individuals that died about 14,700 and 12,200 years ago, respectively. The older one shows that water voles were present in Britain just after the formal end of the last glaciation (more or less 14,800 years ago). The younger specimen is significant because that particular vole lived during a period called the Younger Dryas, which was another mini cycle of re-glaciation and glacial retreat that happened between 12,900 and 11,600 years ago. That period probably wiped out quite a lot of British wildlife populations, but it appears the water voles managed to survive in a few pockets.

The next-oldest samples of water vole remains comprise two skeletons from 4,600 years ago* and another one from 3,000 years ago. These three specimens seem to have a different genetic origin to the older water voles, probably because they came from a second wave of colonizers that scampered across into Britain from Belgium or Germany after the Younger Dryas. This second wave replaced any remnants of the original water voles in England and Wales, but in Scotland those early adopters seem to have had time to re-establish before the new bunch arrived. In modern times there remains a genetic distinction between water voles from England/Wales and those from Scotland, most likely because their ancestors came from different places.

Anyway, to cut a very long story short, water voles have

---

* These were excavated from a Bronze Age barrow. On discovering this snippet I indulged in a happy few minutes of wild speculation about what on earth they were doing in there. Did the vole equivalent of Indiana Jones get squished by the rolling-rock booby trap? Or were a treasured pair of vole pets buried with their master? A tasty vole stew for the afterlife? Then I learnt they were probably living in the cairn and died there, after which their bones fell in. To be honest I prefer my explanations. But in any case, their presence there means they were a maximum of about 4,600 years old or so.

been found across the length of Britain for millennia. In fact, the evidence strongly suggests that they've been in continuous residence here far longer than humans. Our best estimate is that Britain has been continuously occupied by water voles for at least 14,700 years, whereas humans have been continuously present for only 11,600. To put it in one final context, the first reliable historical written records that describe Britain date from 325 years before the birth of Jesus Christ. They came from a Greek navigator called Pytheas. When he got here and put stylus to parchment (or wax tablet, or whatever), water voles had already been in Britain for 12,000 years.

If they don't belong, nothing does.

And in terms of ecological belonging, they are born survivors. Give them water, give them plants, and they're sorted.* In the lowlands, water voles flourish among tall herbaceous plants and grasses – which they eat and hide in – with deep water to dive into and high banks for burrows and shelter from floods. In fens where the banks are low, they can be

---

* There are always a few exceptions to any rule. In our case there are a handful of colonies of fossorial (living underground, away from water) water voles, including one lot in Glasgow, weirdly. If you ask me, they're just showing off.

found nesting in tussock sedges, which are large and fibrous, several feet tall, and resemble Cousin Itt from the Addams Family. (The voles burrow in the bottom and carve a path up inside to a warm, dry nesting chamber near the top. These are great for keeping above winter floodwaters.) In some marshes the voles weave rugby-ball-shaped nests of grass among reeds and rushes, the vole equivalent of a tent.

In lowland habitats water voles live anywhere with good vegetation. Places with closed canopies of trees won't work because the banks beneath are too sparse. But a good few kilometres of open, lush waterway, marsh or fen will provide room for hundreds of voles in the summer. That's a population big enough to weather a few bad years or disastrous floods and come back fighting.

By contrast I have already described the networks of tributaries in moorlands and uplands where a patch of suitable vole habitat might be sufficient only for a family or two, a handful of adults and juveniles. The next patch may be kilometres away, either on the same stream or a risky hike across the moors over to the next headwater. These tiny populations are isolated and inherently vulnerable. A run of bad luck, a flood, a predator, a cold winter – all of these could wipe one out. But there are lots of these patches. And maybe another patch has just had a bumper year. From there voles disperse, heading out from their crowded home to seek their fortunes. If they find an unoccupied patch, it's theirs. And so in these harsh, sparse uplands a population of water voles is a strange thing, spread across tens or hundreds of minute habitat patches. Any of these patches could be occupied or unoccupied as fate dictates. But over time, the whole thing is stable, linked by intrepid dispersers.*

The sheer range of places and lifestyles to which water voles can adapt tells you how hardy, and flexible, they are.†
And they need to be. Because they spend every second of

---

* The name for what I'm describing is a 'metapopulation'. There are some wonderful mathematical models that predict the circumstances under which metapopulations work and fail. You need to know the birth rates, death rates, patch sizes, patch density, distances, dispersal rates and dispersal mortality rates. Then you can use equations to model the 'metapopulation dynamics' and predict outcomes. This is fun if you're a mathematical ecologist. But I'm not.

† I once trapped a good-sized population of water voles that had taken up residence in Reading sewage treatment works. And I heard of some others that had decided to burrow into the thatch of a nearby cottage. (Kenneth Grahame was right. Water voles do indeed break into people's houses.)

every day at risk of being munched to death. They are right at the business end of the food chain. A water vole would be a good meal for anything that preys on the small and squeaky: herons, foxes, stoats, weasels, owls, pike, sometimes even an otter or two. I was once working in a marsh just after a digger had dredged weeds from a ditch and dumped them on the bank. It left one entire side of a two-hundred-metre length of vole-occupied vegetation buried under stacks of gently steaming weed. Perched atop that weed, evenly spaced, and stock-still like participants in a particularly vicious game of musical statues, were twenty grey herons. They knew if they waited long enough a discombobulated water vole would struggle free and make a break for it. That's the mentality water voles are up against. For most predators they are well worth a quick stab. Literally, in the case of herons.

From a vole's perspective, being eaten is pretty much natural causes. But they have a suite of excellent anti-predator strategies. Attacked by a fox? Run for your burrow. Stoats and weasels can follow you in, but your escape tunnels, low down in the bank, water-filled and leading to the river, will save you. How about a heron? Leap for the water, dive, swim down. Chased by a pike? Find a burrow entrance, head up and into your tunnels. Oh, and definitely keep on breeding like there's no tomorrow. Because with your eight-month life expectancy, there's a good chance there won't be.

Water voles breed as long as there are plants to eat. Between April and October, when the vegetation is growing, one female can have six litters of six pups. And pups born in April can have a litter of their own by June. Two water voles can become hundreds in just a few months. Despite the predators, water vole populations quadruple during the breeding season. But when winter comes, breeding ceases, and predation continues. Water vole colonies lose three-quarters of their individuals, which is normal. And when the weather warms, breeding starts again, and the cycle continues.

I can't tell you how much I admire the voles' toughness. They live in constant danger, but spend any time watching one and you'd never know it. I had a Cotswold study site, with a tiny bridge spanning the river Windrush. In nice weather I'd sit on this, dangling my legs and watching the world. It was there, while I was finishing a sandwich, that I spotted a water vole paddling, angled upstream against the current. It was carrying a good chunky length of reed sweet-grass (*Glyceria* *maxima* for the botany buffs)* in its mouth. The water vole took this to the shelter of the bankside, climbed out, then perched on some mud while it scoffed down the stem. It dropped the remnants and indulged in some whisker grooming and a bit of a scratch. And then, for a while, the vole simply sat back on its haunches and regarded the world with me. It seemed peaceful, happy and relaxed. Eventually it dropped to all fours and trotted off to be lost among the reeds.

To me this is what it means to belong: to know what your world is, to accept it and to fit. The reeds, the banks, the depths, the plants, the flow of the water and the way their home can change from welcoming to perilous in an instant, all of this is what makes up life for a water vole. It is their birthright, earned through countless generations of toughing it

---

* If you're into Latin binomials – and, let's face it, who isn't? – it may interest you to know that water voles when I started working on them had the scientific name *Arvicola terrestris*. This has since changed to *Arvicola amphibius* for obscure reasons that a taxonomist would no doubt delight in explaining in excruciating detail. It can cause real conservation problems worldwide when a species that is protected by local laws under one name is renamed as something else. When this happens, or one species gets split into two on genetic evidence, the wild populations can be left unprotected before the law catches up.

out and living in the moment. For fourteen thousand years, predators have feasted. For fourteen thousand years, water voles have ducked, dodged, fed and bred, and for fourteen thousand years their populations have boomed and bust with the turning of the seasons. From the retreat of the last ice age to the end of the Second World War, the length and breadth of mainland Britain has been the epitome of replete rodent-dom. And yet, as I write this, the final few survivors of our national population are hanging on in pockets. Today's voles are the last, tattered remnants of a fourteen-thousand-year heritage.

The first hints that things were going wrong came in the 1980s.* The omens in question were delivered by two reports by the Huntingdonshire Fauna and Flora Society. They showed that water voles had been lost from some places in the local area that had enjoyed long histories of vole-worthiness. So far, so parochial. I was respectively four and six years old when the reports came out, and even then probably could have told you that they were unlikely to set the world alight.†

---

* An observation that applies across many aspects of modern life.

† I'd probably have phrased this as, 'Mummy, who's Flora?'

But actually, to everyone's credit, the findings were taken seriously enough for one of the authors to get funding from Britain's main environmental authority to investigate further.

These further investigations proved tricky. The plan was to compare historical reports of water voles' presence with findings from a contemporaneous national survey of birds on waterways – which also recorded the presence of riparian mammals such as water voles that leave easy-to-spot signs. The snag was that the historical reports weren't great. Before 1965 nobody was systematically recording British mammals, so our best sources of information were hunters' game bags (i.e. their personal records of animals they had shot) and any notes kept by particularly enthusiastic local nature groups or reserves. Game bags weren't a lot of help for water voles, because nobody in their right mind sets out to stalk a rodent. And so for the first half of the twentieth century the only reliable records were from about one hundred to two hundred places that had resident natural-history obsessives. In the 1960s, thanks to the recently formed Mammal Society, this number rose to four hundred locations. And as their recording programme rolled out further in the 1970s, it increased to 1,400.

This caused a problem. The ever-increasing observer effort

meant that the recorded size of the national water vole population increased over exactly the time period for which we were seeking evidence of a decline. Which was unhelpful. To get unbiased data needed a clever approach, albeit from the rickety end of the scientific method. Recorded numbers weren't useful, so instead the researchers examined how reports referred to water voles. For example, before 1930 water voles were described as 'common' in the vast majority (more than 96 per cent) of reports. During the 1930s, about three-quarters of reports said they were common. During the 1960s this was 70 per cent, and in the 1970s, 60 per cent. In the 1980s that national bird survey asked 57 people who regularly visited water vole sites what they thought had happened there. Half of the respondents said numbers had diminished.

Let's be clear: these data are dodgy as heck. But they pointed to a national decline that was gathering pace. By the time I was thirteen years old (in 1989) we had a fair suspicion that British water voles had suffered a long-term decline. But we would need some proper numbers to back this up. And again, to our credit, that's what we got. But this time it wasn't funded by government. Instead the money had to come from an eccentric philanthropist, the late Honourable

Vincent Weir, who ran the Vincent Wildlife Trust.* The first National UK Water Vole Survey was conducted by one man, the gentleman and legend that was Rob Strachan,† working for the VWT from the back of a specially procured camper van. Rob lived in that van for two years, during which he visited three thousand sites across Britain. From the Scottish Highlands to the Chichester plain he recorded the density of water vole sign, details of their river habitats and the

---

* When I say 'eccentric', I'm not kidding. He was, for example, reputedly averse to the colour blue, to the extent that if you wore a blue jumper he'd ignore you until you took it off. But this in no way detracts from my admiration of his intervention, without which much of our subsequent work could never have happened.

† Rob was one of the finest natural historians, and finest people, I have ever been lucky enough to know. Walking with him outdoors meant gaping as his unassuming fieldcraft revealed harvest-mouse nests, badger latrines, deer slots, bat roosts, fox hairs, shrews, great crested newts and God knows what else that we had all obliviously wandered past. He just seemed to attract wildlife. Walk ten minutes with Rob and you could almost guarantee that a troop of something rare would tap-dance across the path in front of you. You couldn't watch nature documentaries with him because he knew more than the narrator did. I so often wanted to accuse him of making stuff up, but every time I was sure I'd got him he would be proved right. Dammit, he was brilliant.

presence of other mammal species. Two thousand sites were randomly selected to create a broad measure of national vole occupation. The other thousand were known to have been historically occupied, and so would give an instant measure of any decline.

Rob's report came out in my penultimate year of school (1993), and it was frightening. Across England, Scotland and Wales (water voles never made it to Ireland), less than half of all locations now had populations. Only a third of sites occupied before the 1940s still were. And the rates of loss were accelerating. The second national survey (conducted by Rob's brother, Chris) was conducted in 1996 to 1998, and by then we'd lost another 68 per cent of the sites that had been occupied in the first survey, six years previously. That's not a decline, that's a calamity. In sixty years between 1939 and 1998, 98.7 per cent of all British water voles vanished. And most of that happened in the 1980s and 1990s.

I was a young child when the Huntingdonshire Fauna and Flora Society published its first concerns. By the time we'd got proper survey data on the scale and extent and rate of change of the problem, I was a grown, professional ecologist. And the situation had deteriorated beyond recognition. The carnage was a world-beater. The rate of loss of British water voles out-stripped that of Sumatran tigers, mountain gorillas and white rhinoceros. And it left us with two obvious questions: what was causing it, and could we persuade it to please stop?

Early attempts to pin down causes form a brilliant example of what happens if you ask a lot of experienced people their opinion when nobody has any actual evidence.* Sure, the real causes of the national vole-vanquish were among the listed suggestions, but so were a lot of things that were nothing whatsoever to do with anything. For example, of 184 early reports, about two-thirds blamed predators, in particular owls, along with stoats and weasels and American mink.

---

* As they say, the plural of 'anecdote' is not 'data'. Just for clarity, by the way, I'm not talking about the equivalent of climate change, in which expert opinion is based on multiple decades of carefully designed, scrutinized, tested and peer-reviewed, independently verifiable and so-far-above-reproach-that-it's-stunning-that-our-governments-still-haven't-done-a-damn-thing-about-it data collection and simulation modelling. That's different. What I'm referring to here is asking some folks to speculate about the possible causes of a brand-new problem without any evidence substantiating their views. It's useful for getting a good candidate list of causes, but not much beyond that.

Other studies blamed habitat destruction, human disturbance, pollution and climate change. These were absolutely the best guesses at the time, but clearly not a good basis for, for example, going out and clobbering the owl population. No owls were harmed, but I suspect that if they had known the accusing finger had been pointed at them, however briefly, they'd have been pretty indignant. (Hard to spot, though, given your typical owl's resting expression of supreme indignation.) An owl might indeed eat one or two water voles a year, but the scale of decline would have required either a radical shift in the owls' diet or their numbers to have somehow increased a hundredfold. The former didn't happen and someone would have noticed the latter.

'Have you seen the heaving plague of owls infesting our garden?' said nobody, ever. It wasn't owls.

Two research papers, published before the first national survey, were pivotal in revealing the true causes. The authors had been studying water voles in a network of rivers in the North Yorkshire Moors. They counted latrines (piles of droppings) to give a very rough approximation of the numbers of water voles at a given site. Of the sites with no voles whatsoever, about half weren't suitable (e.g. under trees). But about a third or so were suitable but were still unoccupied. Unoccupied sites can happen by chance, of course, but not in these quantities. And a major correlate of the absence of water voles was the presence of American mink. In fact, it was a straight line: the more mink, the fewer voles. And during two years the authors recorded mink exterminating water voles from good-quality, high-density vole habitats. When their work was published in 1990, they concluded that mink posed a serious long-term threat to the survival of water voles.

Their conclusion was borne out by the national survey, which did on a national basis what they had done in Yorkshire. The density of water vole feeding sign and latrines across the whole country showed the same, negative relationship with American mink. The evidence was strong enough for the authors to ask (in 1993) why there had been such a 'long debating period' that mink were involved. American mink were destroying our water voles. And there was little suggesting any other cause, except some habitat loss, mostly early on.

To be honest, no other cause was needed. When my predecessor at Oxford began his doctoral studies of water voles in the Upper Thames, his idea was to assess whether habitat quality could serve as any protection from American mink. He got some great data, and showed that good habitat might indeed slow the mink down a bit. But his work nevertheless became a real-time observation of how quickly water voles could be exterminated from a whole river catchment. In 1990 water voles were still common across the region. By 1995 they were restricted to the most upstream reaches of the tributaries, where mink had not yet ventured. And by the time I began my studies, in 1999, they had been all but annihilated.

Which begs the question of how a North American animal came to be rampaging around a country that is separated from its native ecosystem by the Atlantic Ocean. The answers, it won't surprise you to learn, are 'human intervention' and 'profit motive'. In the 1920s there was quite a fashion for fur coats. To feed this fashion some unfortunate American mink were brought over for fur farming. And they got out. (Which is what always happens.) As with all new populations they took a while to establish, but then began to expand exponentially. The rest you know. But as a piece of side conjecture, it's likely that the original escapees didn't establish quickly not only because they were in low numbers but also because they encountered our dominant, native, weasel-type, river-dwelling toothy item. Or 'otters', as they are known. Otters don't like sharing. They are part of the same family (the mustelids) as mink, and it's a family that likes to fight. I've been told of mink found mysteriously killed, the males with their testicles chewed off – an otter's way of hinting they weren't welcome.* Anyway, the suspicion is that while mink were sparse, the otters did a grand job of stopping them getting much of a clawhold. Which makes it even more unfortunate and stupid that we accidentally poisoned the otters by releasing industrial quantities of PCBs and spraying a load of organochlorine pesticides on to our crops. The latter were meant to improve yields in the 1950s and 1960s. They did, but also washed tonnes of toxins into our waters. These bioaccumulated up the food chain and transformed our otters' favourite fish from wholesome to deadly.

The otter population crashed. And with suspiciously correlated timing, the mink population began to properly establish. The mink were blithely unaffected by the pesticides because they have a far more varied and terrestrial diet. And,

---

* *That* never got mentioned in *The Wind in the Willows*.

especially with the otters gone, any fresh fur-farm escapees arrived in Utopia.* The rivers held no predators or competitors. Our waters and their surrounds heaved with naive prey: rabbits, fish, water fowl, eggs and a creature called a water vole. And these water voles were perfect. They looked like the musk rats the mink ate back home but with none of the defences. A straightforward hunt on land, a fun chase in their burrows and easy to catch in the water. Simples (to quote an advertising meerkat).

Water voles' suite of anti-predator adaptations had served them faithfully for fourteen thousand years. But were now worthless. Even their prodigious breeding couldn't save them, because mink have their young in December, right when water voles don't. A nesting female mink can take a vole a day for three months to feed to her kits, and in so doing eradicate a colony that in the summer would be hundreds strong.

---

* Although I should note that in Scotland the otter population didn't crash and mink have still invaded. I could explain this by saying that once mink were firmly established and there were lots of sources of incoming dispersers, the otters simply couldn't keep them at bay. But this is conjecture. We will never be able to prove anything one way or the other.

Zoom out from all of the details of studies, uncertainties, conjecture and discoveries, and the recent history of British water voles is straightforward. There was an initial decline between 1940 and 1950, followed by a catastrophe that kicked off during the 1970s and gathered pace during the 1980s and 1990s which wiped out nearly the entire national population. The initial decline was relatively minor. It was caused by agricultural policies after the Second World War when Britain was seeking to increase its food security. Farming intensified, and to create cropping space, wetlands and ponds were ploughed up. Where once rivers had been braided, wide and marshy, now they became constrained and linearized. Cattle and sheep were piled in increasing numbers into fields. High banks and riparian plants became trampled mud and sludgy shores. For the water voles, which lived in those banks and ate those plants, this was ruinous. They lost a lot of habitat. But the population survived, albeit diminished and battered. And then, twenty years on, the catastrophe proper began. Because the mink had begun to spread.

The water vole's story is partially one of habitat destruction, and perhaps of agricultural chemicals whose environmental impact had not been studied but which were deployed anyway, with a shrug and crossed fingers. But mainly it's the story of people shipping unfortunate creatures across an ocean to farm for their skins, to make money. And who were then careless enough to let that creature out. (Which, as I say, is what always happens.) It's about how, as a result, fourteen thousand years of the presence of a diminutive and glorious component of British wildlife was all but eradicated in seven brutal decades. And about why our lives today are poorer as a result.

I could have allowed the single paragraph above to tell the whole story. Or, conversely, I could have filled this book with many more chapters detailing minutiae of studies and fascinatingly esoteric facts. But I wanted to give enough detail to demonstrate the sheer amount of dedicated and intricate work that is required to uncover the causes of even a relatively simple conservation issue. All those studies, all those years during which I grew from child to adult, represent the very beginning of the process. An idealized conservation project can be thought of as five stages: reveal the decline; confirm it and diagnose the causes; research and test solutions to combat the causes; roll these out, while monitoring carefully, to save the species; then put in place management and more monitoring to make sure the species stays saved. Everything I have written in this chapter gets us to the end of stage two.

Research allowed us to unpick the water vole's story. And it revealed to us what the outcome would be if we did nothing. But if we were to give the tale a twist, and furnish it with a happy ending, we would need far more research still. The job of my Ph.D., and the equivalent projects of my many contemporaries, was to get us through stage three. We would find the solutions. We would delve for data that would form a solid base of evidence, a foundation upon which we could build a national strategy for the water vole's recovery. And that's what we set about doing.

When I stepped blithely into my Ph.D. at Oxford I really hadn't grasped the stakes. I was unaware of the sheer scale of the dedication that had preceded me. I just knew that I was going to be doing conservation, and working out in the field. There was a problem to solve, and I would give it a shot. I reckoned it would probably all turn out OK. And with that sentiment in mind, I hit the field.

There was no way of knowing whether my optimism was well founded. But surely there was no option but to find out. To do otherwise would mean being in possession of a tantalizingly solvable problem and then . . . just doing nothing. To sit back and, despite the urging of all the accumulated evidence, content ourselves with watching as things got steadily and predictably worse.

I mean, who on earth could do that?

3

Volemeister at Large

The legends of fieldwork locate all important sites deep in inaccessible jungles inhabited by fierce beasts and restless natives, and surrounded by miasmas of putrefaction and swarms of tsetse flies.

STEPHEN JAY GOULD,
*Wonderful Life: The Burgess Shale
and the Nature of History*

I'D HEARD THE BURE MARSHES IN NORFOLK WERE GOOD FOR
water voles. So I decided to scope them as a potential study
site. I found a kayak lying in storage at the Zoology department's field station and made some sketchy enquiries, which
(incorrectly) informed me that it had lain there unused for
years. I borrowed it.*

I strapped the kayak to the roof of my car and drove with
it to Norfolk. There I paddled the marsh's backwaters, thoroughly enjoying myself. I wriggled down reed-lined channels.
I skimmed across broads into places where people venture
just once a year to cut the reeds. I found water vole sign
everywhere. Over the next hours I investigated the lakes and

---

* It turns out that it actually belonged to our then deputy head of
department, who used it one week a year for a swan-counting
exercise. He was not at all happy when he found it missing. Sorry.
Again.

dykes and my mind filled with grandiose plans. I would trap reedbeds by boat, or construct floating platforms to survey in places people had never studied voles. It would, I decided, be really cool.

These schemes ended the instant I capsized. Hemmed in a narrow channel by steep banks and tall fringes of reed, I was saved from total submersion only by a desperate clutch at the dyke edge. But the kayak filled with water, dragging me down by the waist. I was, in defiance of common sense – and indeed of all my fieldwork safety protocols – wearing a heavy rucksack rather than a buoyancy aid. So it took ten minutes of silent struggling to heave myself and the kayak out on to something resembling dry land. When I'd calmed down, I cautiously paddled back to the car. I was soaked and chastened. If I'd drowned, and I could have done, nobody would have found me for months. This introduction to the Bure Marshes convinced me of two things: that this was indeed a perfect site for my project, but that I should restrict myself to the bits I could actually walk on.

In April 2000 I returned with a tent, a camping stove and a heart full of hope and expectation. In the boot of my car were boxes of brand-shiny-new folding aluminium water vole traps, as well as one supplementary case of gnawed and battered older traps that had been kicking around a Zoology department storeroom.* It was Monday. It wasn't raining. And I was on a mission to capture water voles – for science, and their ultimate salvation! For six hours I lugged carrots and apples and traps and bedding hay and marker canes up and down the marsh's sinking surfaces. By the end I was aching, sweaty and had set a hundred traps across half a marsh's worth of ditches.† That's two full kilometres of dyke edges. And the next morning I woke early, jumped from my tent, rushed

---

* This time legitimately able to be used. I checked *really* carefully.

† I also, incidentally, learnt the most valuable lesson about working in fenland, which is never, ever, to step on to anything that looks like mud. In many places the 'land' is actually a floating, peaty mat of intertwining common reed rhizomes overlying some pretty deep water. The surface bounces when you walk on it. The paths are littered with black, peaty 'puddles' which are actually holes in the reedmat. These go down a long way into the water below. A misstep could (and frequently did) sink a leg up to the groin. After a week or two, keeping to the green, growing bits becomes second nature. But anyone foolish enough to volunteer on my project was almost guaranteed to forget the initial safety chat and to have to spend at least one morning with a brown leg. Some things you only really learn through experience.

breakfast and by eight was striding out with my vole-handling equipment and clipboard filled with carefully printed data sheets.

This was it. I was going to get my very first proper data.

Except I didn't. Every single trap was still wide open. The carrots and apple I had hidden seductively at the rear of each (cunningly placed just past the treadle that releases the spring-loaded front door) were sitting innocently in place. Sometimes the chunk of apple I'd left outside the front, as an extra enticement, had been filched. But more often it too was untouched. Which is weird, because water voles adore apple.* So my morning ended in disappointment. And when I went out again in the afternoon it was exactly the same story. Water vole sign was everywhere, but I caught nothing. Not good.

---

* They eat 227 types of plant, most of which are green shoots and dry roots with perhaps some tree bark if they fancy something with a bit of flavour. An apple is the vole equivalent of a Michelin-starred dessert course. If you want to go vole watching, and know a place where they live, go when the shadows are long and take an apple. Find fresh feeding sign and leave half the apple on it. Retreat a few metres, sit back and wait. If you're lucky, a water vole will steal up to the apple and sit and eat it. If you're even luckier, it'll try to drag the apple away through the reeds, which never works and looks hilarious.

On Wednesday morning one of the traps was closed. I excitedly picked it up and cracked open the door, to peek inside. I saw fur, and whiskers, and felt tiny feet scrabbling on the trap's metal floor. But the smell wafting out was of fusty cheese, and the creature inside was far too small to be even a juvenile water vole. I upended the contents into my naked hand. Out poured a load of shredded carrot and minute pellets of faeces. Then a terrified field vole* – the tiny, common and (to me, then) completely unwelcome cousin of water voles – plonked on to my palm. It sat and looked at me for a bit, whiskers quivering. Despite myself I smiled (which probably looked terrifying) and gently placed it on the ground so it could get on with its day. Then I cleaned and reset the trap.

That afternoon, when I returned, the same trap was shut. I sighed. Another field vole, more mess. But when I picked it up the trap was excitingly heavy and warm. And inside, curled around itself and desperately hoping I would go away, was a proper water vole.

---

* Field voles really do smell of fusty cheese, by the way, a bit like unwashed feet. Water voles smell different. They are sweeter and muskier.

Oh my goodness. It was months since I'd last handled one. Suddenly I was nervous, and clumsy. My hands shook as I looped the handling net, hand-sewn from a section of fisherman's landing gear, over the outside of the trap. With a gloved finger I held the front door open. The water vole didn't budge. So I cracked open the trap's rear door and blew inside. The vole shot out, straight into the net. I grabbed the material behind it, sealing it in. I got the net off the trap, then tied an overhand knot in it.

I sat back. 'Right. OK,' I breathed. 'Let's get you tagged.'

I took off the gardening glove. The vole occupied itself with attacking the net with its teeth, so I manoeuvred my hand around behind it. I positioned my fingers ready to pinch the vole's scruff. It stiffened as I made contact. I felt its warmth, its beating heart, and the sense of bones beneath its fur. Then I pinched its scruff and held it up to see what sex it was. A male. He froze while I quickly injected the grain-of-rice-sized micro-transponder under the thick skin of his scruff. I put him down, made sure he was OK, scanned the transponder, wrote down the identifying number (BFAF203), weighed him (290g), untied the net and laid it out flat, facing the dyke. After a moment he ran out into the reeds. He crouched there, hidden, for a few beats. Then I heard him slip into the water and paddle away. Done. Good.

I sat there, feeling my pulse and breathing settle. Then I tidied the gear. Everything went back into its compartment in my tackle box, and I washed the trap in the dyke and refilled it with food and bedding. When I stood up it was with renewed optimism. Maybe they'd just needed time to get used to the traps. I fervently hoped so. After all, tagging this vole would have been pretty pointless if he was going to be the only one.

Over the next two days I caught a measly four more. None of them, except that first, was in any of my new traps. For some reason they were all in the small number of battered old ones I'd set. The week ended. I packed up and drove, sulking, to my girlfriend's house for an overdue shower. I brooded for days. What had happened? This was a marsh clearly stuffed full of fluffies, and water voles are not hard to catch. Some textbooks cheerily warn that individuals get so trap-happy they'd get snared every day if they could,* which can stop

---

* This is true. And it can be frustrating. I remember actually shouting, 'Oh, for crying out loud!' (or more sweary equivalent

you from catching more trap-shy individuals. Hah. If only. Perhaps I was just terrible at my job. Possibly true, but why?

The clue was in the field vole. It was not a coincidence that the only one of my new traps that caught a water vole was one that had caught something else first (something a bit, well, dimmer). Water voles may be rodents, and happy to blunder into a box after food, but they're not completely daft. They have an acute sense of smell and know when something isn't right. All my brand-new traps carried a lingering scent of the workshop, of machine oil. I should have cleaned them more thoroughly. Testing this theory, I scrubbed them down then dunked them in a pond overnight. And when I returned to the marshes in May, I caught twenty-three individual voles many times each. Much more like it.

After that false start the project was go. I wanted to study water voles in marshes and on canals to find out in detail how their populations worked, how they spaced themselves, what affected their breeding success, how they moved around, and

---

thereof) when I'd set a water vole loose and a bare twenty seconds later heard the next trap along snap shut. Evidently it had wandered past and thought, 'Oh look, apple!' Not always the brightest, water voles.

whether the form and connectedness of the habitat they lived in had any effect on this. Did they move differently on canals and rivers – which are a straight-line ribbon of habitat – compared to marshes, where intersecting dykes form networks of habitat that could potentially be accessed by a quick hike overland? I was going to find out. The reason this information was needed was that we already pretty much knew that for water voles to survive, the mink had to go. But even if every last mink were trapped and removed – and it wasn't clear then if such a thing was possible – then what? We could be left with swathes of perfect but empty water vole habitat, hundreds, perhaps thousands of kilometres from the nearest viable population. So any plan to restore the species would have to include targeted reintroductions. And this would need information.

Reintroductions are not simple or easy. They sound easy because they involve animals that should be wild, and empty habitat that should be occupied by them. Put one in the other and it's a problem solved, right? But it doesn't work that way. In practice, a vast back-catalogue of research is needed to tell you exactly which aspects of a particular habitat make it suitable for a given species. Get it even slightly wrong and they will probably die – of starvation or exposure or being picked off by predators.

Once you've sorted out what good-quality habitat looks like, you need to know how many animals to release into what habitat size: what age, what sex, what spacing, what time of year; hard release (just letting them go and assuming they will be fine) or soft release (using pens to give them shelter first); whether they need training (e.g. in hunting skills),* whether they are likely to stay put or run miles away, and what makes the difference. We desperately needed this information for any forthcoming reintroduction attempt. And as a start, we needed to have a better idea of how individuals in a natural lowland population† bred and moved in a variety of different habitat types.

Getting this information was the aim of my research. And

---

* Not water voles, obviously – it doesn't take much skill to outwit a blade of grass. But for predators a common problem is that in captivity we can preserve their genes but not the suites of behaviours they need to survive.

† My research was lowland-focused, but running in parallel were many other projects, including an amazing research team at Aberdeen studying Scottish and upland water vole ecology.

of course it meant a load of work. I spent that first year dividing my time between the Bure Marshes (a national nature reserve), a canal which luxuriated in the name 'the Wendover Arm of the Grand Union Canal', and another marshy nature reserve, this time in the south-west, called West Sedgemoor. At each I trapped for one week a month, every month, during the breeding season, April to October. This was a conservation research project so I received a not-exactly-whopping (but mercifully tax-free) stipend of £8,000 a year, and had a minimal research budget. 'Minimal' meant that to run the project I had to beg unsaleable carrots from local wholefood suppliers and use my own car, which lasted a year before dying from overwork. (Nearly killing me in the process – the brakes failed.) I slept in a tent because my project budget could not afford me even one night of B&B a week. I cooked on a camping stove. Originally this was a cheap one powered by a meths burner but later I splashed out on one that ran on petrol and had two rings. I ate a lot of pasta, Cheddar sandwiches (because you can keep Cheddar for ages without a fridge) and supermarket quiche. I lived in wellies and urine-stained clothes.*

---

* Vole urine, in case there was any doubt.

I washed at weekends, and my girlfriend would greet me from the field by directing me firmly towards the shower.

There are drawbacks to this as a way of life. For example, a lot of riparian plants are razor sharp, as are water voles' teeth and most of the external edges of water vole traps. As a result, my hands were a permanent network of cuts at various stages of healing. On the plus side, after a few weeks fresh wounds began closing more quickly as my body adapted to being damaged. Other sources of exsanguination included being attacked by every species of gadfly and mosquito the British Isles can provide. The former mainly comprised nasty, greyish smaller ones with dull red eyes and vicious bigger ones with green eyes.* The mozzies basically came in two fun varieties which I categorized as 'large and stripey' and 'small and oh-my-god-why-is-my-arm-now-the-size-of-a-tree?'. Both thought I was

---

* You get to recognize the type of incoming blood-sucker from its buzz. And I swatted an awful lot of them. I naturally hate to harm any living thing, but I defy even the most committed entomophile to spend a month in a marsh and not end up making an enthusiastic exception for horseflies. Thankfully they are territorial, so at each trap I could swat a few to give me breathing space. It also helps to wear a hat, because they tend to get confused and sit on the brim, probing for non-existent blood vessels with piercing mouthparts. Horrible.

delicious. Between them the bitey insects had most weather conditions covered. Misty, damp conditions meant mosquitoes. Hot, dry and still conditions meant horseflies. Windy weather was blissfully clear of both, but hot and still days were unbearable. The higher the temperature, the faster and more bloodthirsty the horseflies got. On some days even the reed-cutters – folks who are basically made of leather, sinew and toughness – would refuse to go out on the marsh. The flies on those days were so dense that when I reached for a piece of gear I'd end up with a handful of them.

Heat in general was not my friend. At 28 degrees Celsius, when every tree, herb, shrub, grass blade, mud patch and water surface is determinedly filling your atmosphere with vapour, you feel like you're dying in the Amazon. At the Wendover Canal I did get one blissful month of working from the towpath, hiding traps down by the water's edge and luxuriating in being able to wear normal walking boots. Then some kind soul emptied all my traps and threw them into the canal, ignoring the explanatory note I'd put on each. Which meant I had to work on the other side of the canal, away from the public and only accessible by wading across the water. I had to clomp 16 kilometres every day in chest waders. Hot and smelly doesn't begin to cover it. Sometimes I sat in the canal just to cool off.

On the plus side I did develop an amazing suntan. But only on my arms, neck and face. The rest of me stayed alabaster white because in the field you don't wear shorts or take your top off unless you enjoy bleeding and itching. Couple this with the massive calorie-burning opportunities afforded by being on the hoof all day, and I developed a dark brown head and a thin, white body. I basically spent years looking like an unstruck match.

Rain was worse. The lucky swines who conduct research on birds don't go out in the rain, because there's no point. No self-respecting bird is flapping about in a downpour. Mammals, though, especially semi-aquatic ones, don't care. They're out and busy whatever the weather, so mammal researchers have to be too. Which is tedious. Scanning water voles for their micro-transponders means using a reader, which stops working if it gets wet. So you have to put it in a sandwich bag and try to read the display through scuffed, water-speckled plastic. Recording data means writing on paper sheets with a pencil, which, if the sheet is wet enough, stops working like a pencil and starts working like a chisel. And, of course, everything you possess, touch or try to interact with quickly gets filthy and soaked, and causes water to run off you on to everything you're trying to keep dry.

But that's the deal. Come rain or shine, you catch water voles. You set a trap, you have to check it, tag the contents, weigh them, let them go. To do otherwise would be unthinkable, both in terms of the animals' welfare and the integrity of your study.

I'm nearly done, but while I'm complaining, I'd like to point out that the sleeping arrangements were far from ideal. There are many more ways of being awake in a tent than asleep in one. At a campsite this is often because of some ignorant bloke in a string vest and dodgy sandals. Near the Bure Marshes this was because of Fieldwork (the goddess in question having now metaphorically donned a string vest and set about being as ignorant as possible). Sure, waking there was often lovely. Some mornings the lapping encroachment of sunlight was ushered in by rising birdsong, which greeted the delicate dawn with scattered trills and solo virtuosity, combining, swelling, gathering and melding into a liquid, near-infinite, soundscape of glory.* But most of the time my

---

* You've never heard birdsong until you've camped in it. Especially near a marsh. In the spring the noise seems to come from everywhere.

early mornings were somewhere between 'urgh' and 'argh'. The former involved limboing out from a soggy tent flap into greyish drizzle, forcing my feet into damp boots and clumping over to our rudimentary toilet facilities. The latter was the only legitimate response to Fieldwork's boundlessly amusing stock of practical jokes. For example, I once turned over in my sleeping bag to find my face submerged. When I was done coughing up water, I discovered that the tent, my sleeping bag and clothes were all two inches deep. Because an overnight thunderstorm had turned the entire field into a big, shiny lake. And I cannot do justice in words to the primal terror that gripped me, at 2 a.m., when some beast began rasping and huffing around my tent's waterproof exterior. It panted in my ear for a bit then devoured something right next to my head. Daylight and rationality revealed this to have been a

---

It's the only sound I have ever heard that had layers and depth stretching into the distance in every direction (except down, of course). It was staggeringly beautiful. The effect is only slightly ruined by the realization that the messages behind the songs are a combination of 'All this is mine. Don't even think about coming in here, because I'm massive and aggressive, right?' and 'At the same time I am also extremely sexy and very available right now. So how about it?'

hedgehog, snaffling the packet of ham I'd stored under the awning.

Other barriers to slumber included barn owls wheezing and screeching in the roof of the nearby workshops, the roaring of Chinese water deer,* and a weird squeaky grunt I kept hearing above my tent. It approached then receded just once at about 10 p.m. every night. This turned out to belong to a woodcock, patrolling the circumference of its territory at dusk. (The activity apparently is called 'roding'.) But the single most effective sleep disabler I encountered was the stupid bloody pheasants. All that clucking. And flapping. And clucking and flapping and clucking. It started at four and went on and on until I was ready to throttle the lot of them. By June in

---

* These are a smallish, invasive species of deer that have taken to living in the fens of East Anglia, among other places. They are brown, bizarrely tusked and for some reason able to roar like a puma. Which can be unnerving, to say the least. One scared me half to death while I was out alone in a marsh – to the point where I pulled out my penknife in self-defence. When the roaring ceased, a dog-sized specimen trotted round the corner, spotted me and dashed off in a panic. I put my knife away, feeling silly. (PS I learnt while writing this that Chinese water deer tusks are moveable, and have 'combat' and 'grazing' settings. How magnificent is that?)

my first year my mental state had deteriorated to the point where I decided that, yes, I would deal with the pheasants once and for all. I hurled myself into the pre-dawn chill wearing nothing but my underpants. I singled one out for punishment and chased the thing around the field for the five minutes it took for it to remember it could fly. I didn't get close to catching it, but I did feel a lot better.*

I offer the above, and the resultant sleep deprivation, as partial explanation, and mitigating circumstances, for some of the daft things I've done in the field. And I could go on

---

* Yeah. I know. And it came back the next morning. But the award for fieldwork-related strange decisions belongs to my research assistant, Ian. We'd found some funds to get me help, and he'd unwisely agreed to join me in the field. During a particularly frosty April's trapping I couldn't find him anywhere at 7 a.m. when we were meant to be getting up. He wasn't in his tent or at large about the site. I finally found him asleep in our car. His sleeping bag hadn't been up to dealing with the sub-zero temperatures. Freezing, he decided that the car would be warm, and went and sat in it. The car, being made of glass and metal, was not warm, so he put the heating on. Which meant starting the engine. And then, given he was awake and cold and the engine was on, he decided he may as well drive to Great Yarmouth, 20 miles away. He watched the sun rise over the sea, came back and fell asleep. Good work.

complaining, but you get the idea. Field research is a not even slightly comfortable way of spending your time. By now you may be wondering what would possess anyone to do that to themselves, especially for little money and no guarantee whatsoever of success.

In response I can tell you that alkaline fens are perfect breeding grounds for milk parsley, which is the food plant of the caterpillars of the swallowtail butterfly. In May as I walked the mown paths of the Bure Marshes, gorgeous flights of swallowtails would drift into the air and settle as I passed.* Or I could tell you about the barn owl I watched hunting for rodents in the sedge beds early one morning. The weather was stormy, and an orangey sun glared through a break in ominously dark clouds. Against this backdrop the owl glowed stark white as it quartered the landscape. (The weather

---

* They are so beautiful that, in typically disappointing fashion, they have a monetary value, based on the sums that collectors will pay for them. I once met a shifty-looking guy driving a battered pickup truck on the road that ran past the marsh who asked me if I'd seen any swallowtails. I surmised that his enquiry was not for the love of butterfly spotting and conservation. One flew past us as I was busily lying to him that I had never seen any in the marshes, ever. Thankfully he didn't notice.

overnight had been terrible, which was probably why it was now forced to hunt in the daylight.) And I could tell you about how the Wendover Canal had been claimed by vast, floating mats of fool's watercress, having long since fallen into disuse as a thoroughfare. This made it a water vole paradise. On sunny summer evenings they were out, perched on the floating plants and cheekily munching their fill. They were safe from most terrestrial predators and knew it. They didn't give two figs about quiet passers-by on the towpath. So in the late afternoons, on my hike back to the car from setting traps, I would sometimes just sit and watch them from a distance. Or instead I'd spot dabchicks (otherwise known as little grebes) as they dived down and bobbed up like bath toys.

Or I could list the unintended detainees collected by my water vole traps. It comes in at thousands of slugs (which I could have done without),* hundreds of field voles and bank voles (which make a mess, but have the benefit of being

---

* They make a right nuisance of themselves. They slime their way in, bore massive holes in the carrots you've left for the water voles, defecate, then find themselves a cosy spot to pass the time. This is usually right under the treadle, which renders the trap inoperable. The only way of sorting it is to pull out the wire that holds the trap

adorable and able to be handled without gloves), tens of common shrews and several water shrews. Water shrews are cool and lovely creatures with jet-black fur but white bellies and white eyebrows and hairy white feet. They have no fear whatsoever. They are one of the few mammals around with a venomous bite. They bob like corks in the water and swim like they've been wound up, paddling away with their back feet. They were willing to sit in my hand quite happily wiffling about them as I returned them to the bankside. And most people will never, ever, get to see one. I caught a stoat, too. It was surprisingly chilled about the experience of being emptied into a big plastic sack so I could get it out of the trap without being mauled. And also a mole, which ran to the nearest mound of earth, stuck its head into it, then waited like that with its bum in the air until I had gone.

Oh yes. And once I caught a toy rabbit.*

---

together, dismantle everything, hoick the slug into the grass and then spend ages trying to put the wire back.

* There were some volunteer workers on the fen one week and I'd been chatting with them on my rounds. I should have realized something was up when I spotted them expectantly hanging around one of my water vole traps. I said thanks for the rabbit and kept it as a

For me almost any of these experiences (barring the slugs) would justify the discomforts many times over. But the single most wonderful memory I have of any time I have spent in the field is of the dragonflies. I saw them outside the Bure Marshes, among the workshops and fields where I camped. In the yard was a low, plank platform of indeterminate purpose. It probably wasn't used for anything but was too big to get rid of. It smelled of creosote and was rotten in places and I tended to use it as a table for my camping stove. The whole site was bone dry after a long hot spell and the ground around it had started to crack. All afternoon on this particular day, from out of those cracks, flying ants had been struggling free and meandering about in disarray before taking flight. By evening they were everywhere. The earth and sky were filled with them. They were an odd omnipresence: there but just beneath notice. Occupied with cooking my dinner, any glance around would show everything looking much as it always had. But if I concentrated for a moment I could see countless specks

---

memento . . . and only weeks later began to wonder whether it might actually have belonged to one of them, and whether they wanted it back.

weaving in the darkening sky, and across the short-cropped earth thousands more were hustling their unsteady way to where they wanted to be.

I stirred my pasta. A loud, quick clattering made me look up. I glimpsed something slender and blue. It was a dragonfly, and as it flew it caught the evening sun like a brooch pin. I'd never seen one away from the fen itself. It darted back and forth around where I was sitting. So I served up, turned off the stove and settled down to watch. After a moment I realized the dragonfly must be hunting the ants. I could still see their small, chunky black forms hovering around in the dusk. The dragonfly wove among them, as though completing some stupendous dot-to-dot puzzle. If I let my eyes unfocus, there was only the dragonfly, careening across empty space. Blink, and the air thronged with ants again. Another blink and now there were three dragonflies. Then four, all skimming with purpose through the confusion.

Over the next hour, dragonflies multiplied. I had no idea where they came from. They seemed to pop into existence, mid-flight. More and more arrived until there were hundreds. They darted and looped and zipped, moving in one another's spaces. They were all around me. I felt as though I were sitting inside a whirling globe of movement. They seemed to leave bright traces in the air, a lingering glimmer from where their wings and bodies caught the failing light. It was like they were weaving something diaphanous. But nothing here was threatening or claustrophobic, even though they passed close enough for me to feel the air stir. Not one of them touched me. And for my part I did nothing but sit, and gaze.

Eventually the numbers of ants began to thin. The air simply became a little less dense with them. And the dragonflies too began to disappear. The clatter of their wings diminished, and some of the usual twilight noises crept back in their wake. Minutes passed and the field became emptier and more normal. And soon just a couple of stragglers remained, mopping up the very last of the ants. Then they too vanished and it was done. Night fell. I washed the dishes, washed my face and went to find my tent.

I later looked up the dragonflies and found they were migrant hawkers. The species occasionally aggregates in large feeding swarms, and presumably they time their get-togethers to coincide with the emergence of flying ants. Knowing that, and the ecological explanation for their behaviour, however, misses the point. The point is that I once travelled to Peru and saw Machu Picchu as it appeared around a bend in the Inca trail. It glistened like a grey-green jewel on the horizon. I nearly wept.* And I hiked up Huayna Picchu, with its panoramic views down on to the ruins, and photographed a woman doing a headstand above a kilometre straight drop. And I've been to the Salar de Uyuni, Chile, when it was flooded inches deep, dead flat, dead still, stretching on and on into the far distance. The clouds reflected so clearly in that sheet of saltwater that I couldn't tell land from sky. Driving there felt like flying. And I have climbed Mount Olympus, in northern Greece, for days, up through pine forests, with sweeping vistas down to the sea, to the rocky and treacherous summit. Then I hiked down and jumped into the Mediterranean. I'd return to all of these places in a heartbeat. They are stunning, glorious, unique. But there is none I'd choose over sitting on a rotting wood platform in Norfolk, eating a bowl of pasta while immersed in a whirl of ants and dragonflies.

And rather than here, writing this, I would be up a creosoted observation tower overlooking the Bure Marshes. Every

---

* But I had been up since 3 a.m. and was tired from the altitude and a lot of hiking. So probably don't take that too seriously.

horizon here is bounded by trees, deep and verdant. Between them and me lie glistening dykes with their bouncing, single-plank bridges.* Bordered by the dykes are rippling fields of sedge and reed, above which the marsh harriers circle. And overlying and underpinning everything is a strange, busy peace. It's a tranquillity accented, not diminished, by the hum of insects, and the distant shouts and sounds carried from the alien, human world by the odd amplification of flat lands and still waters. These are the places where land and water intersect. The fens are the rich robes worn by lowland rivers. They are places where life thrives. And I love them.

So, overall, as first years of field-based Ph.D.s go, it could

---

* These are known as 'liggers'. The dykes aren't wide, almost (but not quite) narrow enough to jump. They are, however, pretty deep. The liggers are also not wide, and really do bounce. Your first time walking across one is a feat of trust and courage, both in your own balance and in the integrity of their ageing wood. Doing that carrying boxes of traps and bags of carrots seems ridiculous. But after a while you don't even think about it. In three years I never fell in from one, and they never failed. I did, of course, fall in for a large number of other reasons, including because I had stepped on to something that I thought was a deep-rooted tree stump, which turned out to be floating. Life lessons from fieldwork #1: it's never the thing you're worried about that gets you.

certainly have been a lot worse. Despite the false start from my smelly traps, the data were soon rolling in.* I filled spreadsheets whenever I was back in the office for a few snatched days. Life was a mixture of the tediously repetitive, the unbelievably beautiful and the supremely uncomfortable. And that's how it was right up until August. Which was when I encountered the terrible side of working as a conservationist. I arrived at the West Sedgemoor reserve to a strange emptiness. My traps caught half the number of voles I was expecting. The marsh ditches seemed oddly bereft of feeding sign and latrines. I was puzzled, and suspicious. The place should have been heaving as the late summer swelled water vole numbers but instead they seemed to be dwindling. And when I returned there in September, vole numbers were lower still. Finally, when I picked up a closed trap to find it

---

* If you are a grammar aficionado, you'll notice that I use 'data' in the plural. This is because I always got told off for using it as a grouping noun, when it's actually a plural of the word 'datum'. To drum this point into me, my boss, Professor David Macdonald, made me write out 'Each datum begat another until there were many reams of data', or something like that. It was a scarring experience that has left me a terrible pedant.

disconcertingly light, I discovered why. Inside were water vole droppings, and carrot that had clearly been gnawed. The bedding was soaked and sticky. I reached in and pulled some out. It was wet, not with rain, or urine, but with blood. And down the inside of the trap were claw marks, as of a rodent that had been dragged out, scrabbling and fighting for purchase. And further down the dyke, near to a stack of browning feeding sign, I found a small pile of entrails and fur. The fur was instantly recognizable. It had once belonged to a water vole. I stared at these remains for a while before quietly completing my round. I let the remaining voles go with no more measurements or tagging. Then I shut the traps, packed up and told the guys at the reserve that a mink was eating their water voles.

That was the end of trapping at that site. The network of marsh ditches had, to some extent, protected the population from being annihilated. On a river, for example, a mink would simply have worked its way down the population, and there would be nowhere else for new water voles to come from. Empty habitat would stay empty. Here at least the difficulties of hunting a complex of ditches meant that the mink had missed some patches of voles that could recolonize the now empty areas. But in the long term, if the population were to survive, the voles would clearly need further protection. The reserve staff would need to get busy, catching and removing the mink. This was not something I could do for them. I lived three hours away, in Oxford. And for the sake of my study I now had to leave the site. The arrival of the mink had invalidated all of the data I had collected there. It couldn't serve as a representation of how a natural water vole population functioned, and so for my project it had become useless.

Worse followed in the next year. An outbreak of foot and mouth disease among British cattle populations in 2001 effectively closed the countryside to people, including wildlife researchers. I passed two long, under-occupied and anxious months of my field season staring at the wall of my office. By the time I was permitted back to see how my voles had done at the Bure Marshes and Wendover Canal, it was already too late for me to assess which had survived the winter at my sites, and what were the correlates of that survival. Their populations had moved on, with fresh births and months of new deaths since the cold weather. In response I upped the number of trapping periods to make sure the rest of my data were as good as they could be. Then, in October, during my last week of the season, someone I knew at Wendover came rushing over to tell me she'd spotted a mink. Heart sinking, I asked if she was certain. She said she was. So I phoned the body responsible for managing the canal. They managed to source a single mink trap, which I helped them set. For two weeks they checked it and it caught nothing. Then they took it away.

It was at least partially my fault. I hadn't quite believed the person when she told me about the mink sighting. People get these things wrong the whole time,* so I was happy to trust the evidence, or lack of it, provided by the stubbornly empty mink trap. Besides, I didn't have the remit, funds or time to conduct a lengthy spell of mink monitoring on behalf of the canal's owners. It was their job, not mine.

While true, this excuse didn't make me feel any better when I returned in April to find the Wendover Canal emptied of water voles. Four kilometres of bustling irises, sweet-grass,

---

* For example, the vast majority of sightings of a wild 'big cat' in the UK turns out to be a black Labrador or a normal-sized cat that happened to *look* big because the photo was taken from a bad angle by someone with no idea about how perspective works. People love to imagine predators where they don't exist any more.

fool's watercress and reed canary grass now supported a grand total of two individuals, stranded miles apart. There was no nearby habitat full of voles that could come running in to the rescue. There was no saving grace. Thanks to my inattention, and thanks to the actions of one American mink working its way through the voles over the course of one winter, a population I had studied for two years had vanished. Permanently.

I managed to find a replacement site for my final year, down on the Kennet and Avon Canal. I worked hard to secure enough captures and measurements to allow me to write up my thesis. And when my fieldwork was done, I hit the office to analyse my data and to start wrangling my conclusions into some sort of shape. But driving everything, now more than ever, was the feeling that I had something to compensate for.

All that work. All those captures, weighings, taggings and measurements of vegetation. All those voles I had put through the stress and indignity of being my research subjects. All of it, now, had damn well better add up to something worth knowing.

4

A Scrap
of Voles

an unkindness of ravens
a clattering of choughs
a murmuration of starlings
a charm of goldfinches

The Book of Saint Albans, 1486

a destruction of cats
a troop of baboons
a barrel of monkeys
a tower of giraffes

Merriam-Webster Dictionaries

a scrap of voles

TOM MOORHOUSE,
hoping it'll stick

I HAVE TRIED AND FAILED TO FIND RELIABLE EVIDENCE OF A grouping noun for water voles.* This doesn't surprise me. Water voles don't group. Some mammal species live in large social units with intricate hierarchies and carefully curated interpersonal relationships. Water voles want everyone to go away and leave them alone. If it weren't for the imperative to breed, and perhaps a vague worry about where everyone else had gone, they would probably be perfectly happy never seeing another vole their entire lives.

The water voles' approach to life is pretty typical among small mammals.† Female small mammals are often territorial,

---

* I did find some unreliable suggestions, mainly online. Oh deary me. As an aside, though, a friend of mine once told me that a group of dormice is called a 'cuddle'. I so want this to be true.

† 'Small mammal' is a semi-technical term ecologists use, mainly to describe some, but not all, rodents and also other creatures like

meaning they maintain a patch of habitat for their own exclusive use, if necessary through the liberal employment of threats and violence. The strategy makes sense. As a small mammal your life expectancy and ability to breed depend more on the availability of food and shelter than on the likelihood of neighbours leaping to your defence. A female water vole, for example, would definitely not rush to assist if another was being attacked by a fox. First, it wouldn't work, because any attempt to club together would result in a bemused but happy predator merrily scoffing the lot of them. And secondly, well, why risk it? A neighbour being digested represents a prime opportunity to grab some of her territory. I suspect that the reaction of most water voles to hearing of another's demise would be, 'Oh, good!' This doesn't make them heartless, just realists. It's a very tough world for the small and squeaky.

What this all means in practical terms is that for female

_____

shrews that may have adopted one of a particularly non-social range of approaches to life. Just don't ask questions like 'How small, exactly?' and 'Where is the cut-off with medium-sized mammals?' because a) there are no answers and b) everyone knows what a small mammal is, so stop causing trouble, OK?

water voles to stay safe and raise babies they want to own a quality length of river, canal, lake, pond or dyke edge. The territory gives them everything they need, but comes at a price. Because if it is worth having, all the other females will want it too. So it must be defended. Females spend much of their time patrolling their territories, looking for intruders to beat up and topping up the various markers they leave as 'Keep Out' signs. These markers comprise latrines and feeding sign left at regular intervals down the banks, and especially at territorial borders. The latrines are piles of a female's own droppings, left right at the water's edge. She continually adds to these, stamping the droppings flat with her hind feet and smearing them with scent from glands on her flanks.* They stake her claim, and the freshness of the scent tells prospective incomers to expect hostility. A pile of feeding sign might be a consequence of having to dispose of uneaten bits of plant, but probably also has some role in territorial marking.

---

* I am eternally grateful that humans don't have to do this to keep other people from breaking into our houses. Although, thinking about it, a giant pile of faeces outside your door probably would do the trick.

All of this was already known about water voles at the time when the trapping data from my own studies began to roll in, but my results provided a nice demonstration. Very occasionally I would capture two different females in the same trap in the same location, but on different days and only if the trap lay squarely between their territories. I almost never caught an adult female inside another's range. The traps were 15 metres apart, and such a large infraction would almost certainly result in a confrontation. This would involve a lot of splashing and indignant squeaking, and some chasing around the plants, until one of them backed down. What was new from my data was what I discovered about the male voles.

Male and female mammals in general have different priorities. Everyone needs food and shelter, so that's a given. But for many males a successful life is one in which they have impregnated as many females as they can possibly get their paws on. The general assumption for male water voles was that they too kept territories, to secure an exclusive batch of lady voles. As it turns out, though, they don't. The males I caught were happy* to overlap their ranges, and I often caught

---

* OK, not happy. Grumpy and combative. But they do it.

one a few traps deep into another's range. They are not territorial probably because females are strung out along the length of the waterways, rather than forming defensible clumps. Imagine attempting to chat up everyone in a long queue outside a nightclub, while simultaneously trying to prevent anyone else from talking to them, and you'll see why the water voles don't bother. Instead they have large home ranges that overlap with those of a number of females and other males, so they can roam around and keep an eye on which females are coming into oestrus. Then they can fight for access to her. Of course, all the other males are doing the same thing, and so a receptive female quickly finds herself at the centre of some macho vole-squabbling.

And such, for want of a better word, is water vole society. Females are narky and territorial, males are slightly less narky and not territorial. And this antisocial set of arrangements has a whole set of knock-on consequences. For a start it means that the size of a female's territory varies with the time of year. Vole densities are at their lowest in March because of overwinter losses. At the start of breeding season there may only be a couple of adults occupying every hundred metres or so. But then breeding kicks off. By late March, early youngsters are born. By May and June these have grown into adults, are fighting for territories of their own and having their own babies.* By September the same hundred metres of river is stuffed with two or three adult females, the same number of adult males, and loads of babies and subadults. And because the females remain stubbornly territorial the only possible way to accommodate everyone is if territory sizes shrink. They can end up about 20 to 50 metres in length. But as our trapping results showed, bigger females have an advantage. Female body weight was a great predictor of territory size, presumably because heavy females can throw their weight around to get hold of more stuff.

The above all makes behavioural-ecological sense. What doesn't make so much sense is what happens to the males. With their greater flexibility and ongoing quest† for females,

---

* There's an episode of *Star Trek* where the *Enterprise* gets invaded by small, fluffy, round things called Tribbles. These multiply exponentially in cupboards before bursting out and submerging the whole crew in cuteness. That's not dissimilar to a riverbank during water vole breeding season.

† With apologies to Evelyn Waugh and fans of *Scoop*, questing voles are actually not all that feather-footed. They move in a kind of rapid waddle-bustle. But the fens can be quite plashy, so that's OK.

their ranges are much larger than the average female's territory, and because the males overlap with both sexes, you would predict that their range should stay large regardless of the numbers of other voles. After all, if females start bunching together, then keeping the same size range means you just get access to more of them as the year progresses. Which is surely good. But that's not what happens. As vole numbers go up, males' ranges, too, shrink.

It turns out that being able to overlap with other voles is not the same thing as being able to go anywhere without consequences. There is a missing ingredient. Water voles might be antisocial, but they still need to be on good terms with their neighbours. And any male venturing beyond his normal haunts is asking for trouble. Every new female he meets will be deeply suspicious, and probably violent, because he is a risk to her babies. He might kill them to bring her back into oestrus. He wouldn't risk killing his own babies, of course, and so he's not a threat to a female he's copulated with. (Taking advantage of this, females probably copulate with a large number of males to keep them guessing. The suckers have no way of knowing if the pups are theirs, but have to leave them alone just to be on the safe side.) A roving male water vole will also attract aggression from new males because he's an unknown quantity, and potentially going to get between them and their ability to breed. Whereas in his usual patch all the males will largely have already sorted out who can beat whom in a fight, here he's likely to have to prove his toughness.

Male water voles do not live a carefree and feckless life. They are embedded in such a tangle of social interactions that overlapping ranges with more than a given number of other voles comes at a cost in terms of elevated aggression. Again we found from our trapping data that the biggest, heaviest males had the largest ranges, and overlapped with more females and other males. Presumably being large means they can bully a path to where they want to go and then scuffle their way to more one-night stands. But even the burliest vole can only get into so many fights. At some point all males hit an upper limit to the size of their ranges and then stick with maintaining their established relationships.

Phew. So our studies showed that both sexes have to restrict their range sizes as densities increase, probably because of aggression from other voles. But as ever in ecology, other explanations could work. For example, more vole food is available later in the year. And that could result in smaller ranges. Why roam further if you can meet all your dietary requirements on your doorstep? In some places the river-banks are 5 metres thick in food plants, and the river itself is filled with comestible goodies like watercress and irises. Even the shortest territory could yield something like half a square kilometre of edible plants. That's surely enough to stuff a 300-gram rodent and family until they're all approximately spherical.

But . . . there's a problem, which is this: why on earth, if they are all happily living in an emporium of food, are females always trying to grab more space? Well, one good reason is that if they get spare habitat then their daughters can inherit it – a good idea to hand benefits down the generations. There's another answer, though, which lies in understanding a pretty fundamental aspect of small mammal ecology, a topic that my research has contributed to, and which is pretty cool, if I do say so myself.* The fundamental aspect in question is what small mammals do when you give

---

* Although, to be fair, what most academics consider cool is more usually described as 'Seriously? What a geek!' by everyone else.

them food. Which is they make babies. (You can try a fun experiment at home if you live in a town with rats. Simply maintain a few good handfuls of breakfast cereal in accessible outdoor locations and see how long it takes for your neighbours to move out.) Lots of food means more energy for more litters of more young, all of which can be bigger and healthier and more likely to survive. Boom time. And when food starts running out, then females have fewer, smaller litters of lower-weight babies. So the whole system is governed by a feedback mechanism. The rate of breeding goes up until numbers reach the habitat's 'carrying capacity' – the maximum number of a given species it can support – at which point the rate of breeding drops again. But, of course, water voles appear very unlikely ever to run out of food. So such a feedback mechanism shouldn't apply.

Here's where it gets nerdy. From our trapping data we were able to repeatedly weigh a load of individuals over various breeding seasons. By a load I mean that over eight years we caught 1,200 water voles 4,744 times (or about four times each) across eleven sites. We used these data to work out how fast the voles were growing, and compared this to things like range sizes, population densities and measures of plant (i.e. food) abundance.* We discovered that pups in small territories got bigger more slowly than those in large territories. We were also able to show that water voles don't become sexually mature at a given age but at a given body weight. Wherever the voles were, and whatever the time of year, males had descended testes and females had perforate vaginas at exactly the same weight. (Yes, I did have to judge these myself on every single vole I caught. No, it was not glamorous – for either of us.) Any variation in growth rates therefore translates into a difference in how long it takes pups to become adults. In the shortest territories females took a full extra week (seven weeks, not six). And males took an extra five days. So it's a feedback mechanism after all: voles breed

---

* This isn't actually very easy. Growth rates vary with body size: when we are new-born we grow slowly at first, then we start shooting up until we're nearly adult, at which point our growth rate slows to pretty much zero. So we needed to calculate the overall growth curve for voles within each population, and what growth rate would be expected at a given size, and compare this with the environmental factors of interest. I'm not a brilliant mathematician, and these calculations involved what our group statistician likes to call 'doing hard sums'. So I drafted in my brother to help, because he used to be a physicist and apparently enjoys this sort of thing.

quickly, numbers build up, territories get shorter, voles start maturing more slowly.

This finding implies that pups in smaller territories were short on food, hence growing more slowly. But then what about that abundant perfectly edible vegetation I was talking about? Well, there's a catch. Feeding 5 metres from the water is spectacularly risky. It is much, much safer to stay right at the water's edge, ready to jump in at the first sign of bother. Water voles handily leave piles of sign where they've been eating, so we can see this for ourselves. I measured well over eight hundred piles of feeding sign, and all but ten were within a metre of the water's edge. That's where they eat. And a quick recalculation of the amount of food available, limited to just this metre, shows us that small territories actually hold three times less food than large territories. Despite living in a paradise of food, water voles are unwilling to risk touching most of it.

All of this intricate work finally yielded a plausible explanation for why females are always after more territory. It's because then she and her offspring will have more food, and be more likely to get big, be healthy and survive.

At some point, inching closer to understanding my study

animal, I found that I could no longer look at a river in quite the same way. It's not as simple, or as peaceful, as it used to be. To zoom out from a water vole-occupied bank and observe the carryings-on is to witness a constant, all-in rolling squabble over who has the biggest territory, who's trespassing where, and whose babies are these anyway? The biggest females hold the biggest territories and copulate with a number of males to keep them guessing; the burliest males will jostle their way to overlapping with more females and so give themselves the best chance of getting to a female when she's in oestrus. All of them are striving to survive and to maximize the number of their own babies that make it to adulthood. And all of them are up for a fight, if necessary.

None of this is probably anything like how most people perceive something as outwardly sweet as a water vole. I mean, the most common question I get asked about them is whether you could keep one as a pet.* But knowing any of

---

* Folks can take quite a lot of convincing that it would be a terrible idea. Aside from ethical queasiness about capturing and keeping truly wild animals for our own amusement, all of our traditional companion animals are descended from species that have social systems and dominance hierarchies into which humans fit. I mean,

this hasn't made me love them any less. It has made me admire them more. Their innocent, fluffy exteriors conceal dysfunctional social relationships, unexpected toughness, cantankerous combativeness, lots of discreetly hidden sex and a strong desire to be left alone. Kenneth Grahame was right to call them 'ratty'. He was also right that they are quintessentially British, if only in terms of their capacity for insular bloody-mindedness. But otherwise they are nothing like their comfortable literary namesake. They owe far more allegiance to the Wild Wood. And to want to tame one is to entirely miss the point. They don't think the way we do, or act in accordance with any of the values we hold in high regard. They are 'other'. And should always remain so. We may, if we're lucky, sometimes be granted the tiniest peek into their lives: a flash of groomed whiskers, a rustle in the undergrowth, a furry shape industriously ferrying a chosen stem somewhere for its own purposes. And how much more

---

can you imagine trying to impose our behavioural conventions on to water voles? They are genetically predisposed to view us either as a predator or as a competitor. In either case all we can expect is biting. And, of course, latrines on the carpet.

thrilling, knowing that this is a world that has precious little to do with us.

I still don't know what the grouping noun for water voles is. It's probably a herd. But I'd prefer, and would like to propose, a 'scrap'.

Beyond the personal and academic fascination, these details of water voles' behavioural ecology have very practical applications. They are vital to know if you want to release populations to restock environments from which they have vanished. The information would prevent you, for example, from sticking two unfamiliar females into the same release pen (resulting in one victorious, if battered, female when you returned the next morning) and tell you that pens in good habitat should be about 40 to 50 metres apart, so the occupants can space themselves correctly. It tells us that 'good' habitats are those with thick enough vegetation to provide good food, but that there may be an upper limit to the width of vegetation required (although the thickness of the vegetation may also protect from predation). All this is useful stuff, derived from trapping studies. But trapping alone can only provide so many insights. As a tool to study survival rates and sexes and weights, it's great, but it only gives a very coarse measure of how animals move around. If you want more detail about individuals' movements and interactions, and to be sure that you have understood things correctly, you need to be able to get a location for each individual far more frequently than just once a day. And you need to employ a very different method: radio-tracking.

Until recently any documentary about field ecologists working in African savannahs had to include a shot of a researcher in combats and a cotton shirt posing on a Land Rover.* They hold what looks to be a TV aerial. Slung across their shoulders is a receiver, which they adjust, expertly. They wave the aerial around and a beeping noise gets stronger or weaker, depending on whether it's pointing in the right direction. What these researchers are doing is radio-tracking, mostly some big cat or other. Even in the present era of GPS collars (some of which automatically upload coordinates for an animal on a second-by-second basis to a satellite, which relays them down to the researcher's laptop in the comfort of a nice campsite or hotel

---

* Usually gazing away into the sunset. Nobody knows why. It's probably in the contract.

somewhere), these shots still give me research envy.* Radio-tracking big cats makes wildlife researchers look cool. I mean, they have attached radio-transmitters to lions. They're tracking flipping lions! That's cool.†

My equivalent sees me on my hands and knees beside a black, peaty splat of a hole I've dug at the base of a massive alder. And I'm increasingly sure that I've been pursuing an ex-vole. Water voles almost never stray from the dyke edges, and this one is tens of metres into the carr woodland. The beep from the receiver is clearly coming from deep among some tree roots. Damn. I was hoping to avoid this. I can't possibly dig up a tree but I'm not abandoning the radio-collar.

---

* GPS collars still aren't small enough to go on something like a water vole, and if they were would probably a) run out of battery pretty quickly and b) have difficulty contacting the satellite from deep inside a burrow. So the old-school methods still apply.

† To give you an idea of the range of animals you can track, one of my friends works on penguins and has got data on their dive profiles. (Top tip: avoid their flippers, which are positioned at crotch height and capable of administering a life-altering whack.) Another spent a fun night radio-tracking a wood mouse. He checked on it every ten minutes for three hours as it bounced around a tiny area. Then it abruptly disappeared into the high branches of a tree. Owl-assisted, we suspect.

They're a hundred quid each. So I take off my fleece, roll up my sleeve and lie face down in the mud. Then I gingerly insert my hand into the open entrance of the burrow.

This, I should emphasize, was not an intelligent move. The burrow was being used by a stoat. Which meant I stood a good chance of withdrawing my hand to find the stoat attached to my fingers. As it was, I was lucky. The burrow was smooth, dry to the touch and vacated. A short way inside, an opening led into a chamber the size of my fist. It was warm, and filled with what turned out to be fur and soft grasses. I pushed an inch further down the tunnel to find another chamber, cooler, and filled with delicately smooth objects. Bones, and what felt like the shape of my radio-collar. I pinched it between two straight fingers and drew it out with a grin. Actually that was cool. A tidy-minded stoat, with nothing wasted. Eat the vole, and the bones go into one chamber and the fur into another to keep you cosy. I felt sorry to have disturbed its home and did my best to put everything back how it was. But it was worth the hassle to get the collar back.

Radio-collars are built to be tiny but solid and user-refurbishable. They consist of a cable tie covered with shrink-wrap tubing holding a small circuit board and a largish watch battery, all potted in dental acrylic to keep the water out. They weigh 4.5 grams, a tiny fraction of an adult vole's body weight. They are delivered with a set frequency, a tiny whip aerial (which the voles always chew off in the first week, but, hey, it's the thought that counts), and two tiny wires poking out from the acrylic. These need to be soldered together to activate the collar, which has a 5–8-month battery life. Once you've soldered up a bunch and mixed up more of the dental acrylic to dab over the wires, you've done the easy bit. Getting the dratted thing on to an actual vole is a whole different mission.

Step one: get someone to hold the vole such that only the head is poking out of their (heavily gloved) hands. Step two: wait patiently for the vole to stop biting said person's glove, so you have some hope of getting near its neck. Step three: quickly loop the collar over the vole's head, keeping fingers clear of massive and gnashing bright orange teeth. Step four: place a little finger gently between the cable tie and the vole, and tighten just enough that the collar won't pull over the vole's head. Avoid constricting anything vital to the normal operation of the vole. Step five: snip off excess cable tie, glue end in place, release the water vole. Step six: disinfect and bind any wounds you may have received during previous steps.

Many collars slipped off in a matter of hours, because of the need to err on the side of keeping them loose. These, of course, had to be retrieved and refurbished by replacing the cable tie and heat-shrink tubing. Then we had to get them back on a vole. Once we had collared enough individuals, we gave them a week to settle, and went out tracking. Tracking at a site required two full weeks every month in which two researchers took it in turns on a one-hour-on, one-hour-off basis to hike four-kilometre circuits. During each circuit we recorded the geographical location of up to fifteen voles down to the nearest metre. This had to be done without getting close enough to scare them away, which results in a splash, then rapidly receding beeps from the receiver as the vole scarpers. This would mean that you had stuffed up the very thing you were trying to measure. So we got good at triangulating from a distance. From time to time we'd find a vole had left the immediate edge of the water. If they stayed away for more than a day, they had probably passed through the digestive system of something predatory. And so again the collar would need retrieving from whatever weird location it ended up in. I have found a collar deep in the woodland beneath a heronry (I can only assume it caused the heron some discomfort), on a vole floating face-down in a dyke with a hole in the skull and – oddly – no legs (probably whacked by a heron, escaped, died and then got chewed by a pike), and on a vole that was lying apparently untouched right in the middle of a recently mown fenland compartment (maybe dropped by a bird of prey or hit by a mower, but, frankly, who knows), as well as down countless weasel and stoat holes.

Aside from exciting and varied insights into the last resting places of water voles, and a new appreciation for just how tough their lives can be, radio-tracking provided us with a far more detailed view of how they move around. In particular, it showed us that females' territories do not occupy a constant position. Instead, they exhibit what is known as 'drifting territoriality'. To understand this, imagine radio-tracking an animal of some little understood species. As you gain more locations for it, over hours, days or weeks, the observed territory size gets bigger until you hit the furthest extent of the animal's normal movements. At that point more fixes make no difference. For water voles, though, this never happens. The more fixes you get, the bigger the territory gets. And it just keeps on going, for ever. Because the size and location of females' territories are continually shifting.

Partly this is because females die a lot. Any suddenly vacated riverbank will become another vole's smug extension quicker than you can say, 'Should have spotted that weasel.' But females can also lose territory while on maternal duties. Water vole pups are altricial, meaning that unlike wildebeest or antelope calves, which can run almost as soon as they are born, they are completely defenceless. They start out tiny, pink and hairless and, if I'm brutally honest, not very attractive. And until they are up and running (in a few weeks, by which time they are furry and very sweet), they can't be left alone in the nest. So water vole mothers have to forsake territorial duties until their young are mobile. We often saw this while out tracking, with radio-collared females holed up, unmoving, for a full week, often inside a tussock sedge. During this time she would probably only venture far enough from her nest to feed.* I once found such a female accompanied by a collared

---

* In places where voles live in riverbanks you can find grazed 'lawns' around water vole holes, especially those high up in the bank. These may well be formed by nursing mothers grabbing a quick bite.

male, and he stayed with her in the tussock for four days. I still have no real idea what they were up to, and didn't want to disturb them in case I caused the loss of a litter. The demands of data collection meant that I had to move sites before the couple split up. And so, frustratingly, I received this one insight into the lives of two tagged water voles without ever knowing what the behaviour meant, whether it was important or just a one-off. As I say, frustrating. But I digress.

With nursing females off territorial duty, their neighbours have a prime opportunity to claw back some real estate. Of course, once those females themselves have nestlings, then the others can capitalize on their absence, and so on. Through a constant gaining and erosion of lengths of habitat, the exact geographical location of females' territories drifts. This happens particularly because many lowland water vole habitats can be quite extensive and homogeneous. It may not matter too much where a given range or territory is situated as long as it's big enough.

Discovering drifting territoriality in our water voles was only possible due to radio-tracking. And radio-tracking has also proved invaluable for detecting and recording dispersal movements, in which individuals up sticks and hike off to seek their fortunes elsewhere. Knowing how often and how far the voles can disperse is useful if you want an idea of how likely, for example, a population is to be able to colonize unoccupied habitat a few kilometres further down a river. Without a collar, this information would require you to set traps in every last bit of habitat where a vole could conceivably arrive. While this is possible (although a lot of hard work) in patchy upland habitats – where one study showed an intrepid vole to have hiked 5 kilometres in a few weeks – it is not possible in large marshes. There the only real option is radio-tracking.

We derived one more major finding from our radio-tracking. But this one was devastating. I uncovered it during the winter when running statistical analyses on birth rates at our study sites. And when I was sure I had it right, I ran straight out of the office to tell my supervisor. Because what I had found was that the number of females being born into study sites where we were radio-tracking had dropped by a half. Not only that, but the drop occurred only in the sites where we were tracking. Trapping didn't cause it. And the birth rates of male water voles were unaffected in all sites, no matter what. In the radio-tracked sites only half as many females were being born. And that, in a species of conservation concern, is not good.

I scrutinized the figures. I went through them in obsessive detail. And everything pointed to a terrible correlation: the number of females being born had dropped by 48 per cent, horribly similar to the percentage of adult females (49 per cent) that we had collared on each site. This could occur if every collared female stopped giving birth to daughters. But why would that happen? I hit the library, trying to find any previous study that could explain things. Eventually I found two, one of which was in Russian.* The papers showed that water vole mothers that were food-deprived during pregnancy raised litters that were heavily biased towards males. The mothers also had elevated corticosterone levels, a sign that they were physiologically stressed.

These sex-ratio adjustments by water voles make ecological sense. In times of plenty, daughters inherit territory from

---

* In the days before Google Translate, and before academic journals were online, I had to pay to get one key article interlibrary-loaned (it was a study in the journal *Zoologichesky Zhurnal* by G. Nazarova, since you ask) and photocopied, and then pay again to have it translated from Russian into English. It was worth it, though.

their mother. So all being well, it's better to have daughters who will benefit from good food and plentiful space. But when food grows tight, it probably means that your territory is oversubscribed or of poor quality. And so it's better to have sons, because they will head off away from their natal territory to breed (quite reasonably not wanting to have sex with their mother or sisters). It seemed that our radio-collars caused stress to the water vole mothers and the above mechanism kicked in, making them preferentially give birth to males – or perhaps making them wean only their male offspring, leaving the female pups to die. Whatever the exact explanation, it was not something we wanted to have been responsible for. We could plead that radio-collaring is a tried-and-tested technique, that previous researchers had radio-tracked water voles and found no ill effects. (These researchers had recorded only the number of offspring born, though, not their sex ratios.) But while we had no way of knowing that any of this would happen, it doesn't alter the fact that it did.

We stopped radio-tracking before permanent harm could occur. But many studies, including most reintroductions of animals to the wild, will include a budget for radio-tracking, to monitor the reintroduction's success. I am certain that for most species radio-tracking has little impact. But our findings showed that radio-collars have a severe effect on water voles. The last thing you'd want to do, given that reintroduced populations start small anyway, is put a collar on a vole.

The interaction between research, knowledge, animals' conservation status and their individual well-being is extremely complicated. Our actions with respect to all these must be carefully balanced and nuanced. Wildlife researchers have a duty to safeguard the animals they study, and to make the best conservation decisions based on the best evidence. Part of this duty extends to taking a deep breath and holding up our hands when something unexpected happens. We must be willing to report not only the positive results, but also the cautionary tales. This way others can learn from our mistakes. And so we wrote up the radio-collar findings with our explanation, and it was published in a good academic journal.

What happened next was that the journal decided that the results were interesting enough to publicize. A number of media outlets got interested. I was interviewed by Eddie Mair on Radio 4. It was exciting to explain what we were doing, what we had found, and how it meant we should be very careful with even commonly used monitoring methods. It was all reassuringly nice. But a couple of hours later I found myself on a local Oxford radio station, being interrogated by a DJ. The gist of his questioning amounted to, 'What sort of monster are you?' and 'Why did you do that to those poor voles?' Ouch. To cap off a stressful week, a mainstream broadsheet newspaper – which shall remain nameless – ran an article entitled something along the lines of, 'Scientists studying the causes of the water vole's decline are to blame'. The clear implication of the headline was that without scientists getting involved, water voles would be doing just fine. Which isn't true, but hey. And unless the subeditor in question was knowingly referring to a decline of only one sex, at two study sites, in the short term, by accident, and which self-rectified once the voles' collars were removed, then frankly the headline was pretty misleading. The article itself was balanced, but the caption made me fantastically unpopular for a week or two with everyone involved with water vole conservation (many of whom received flak as a result – sorry, guys). The newspaper's subsequent correction was written in a tiny font somewhere on one of the back pages.

Honestly, it could have been a lot worse. That was my brush with the dark side of wildlife science writing. But it can get a lot more defamatory if you work on something bigger and more charismatic than a water vole. As species get rarer and more iconic, the emotiveness of the surrounding human issues also increases. By the time you're working on big cats, for example, conservation is hugely political and opinions plentiful and polarized. These situations breed the type of person who meets reasoned debate with inflexible positions and ad hominem attacks. And, of course, some media outlets are willing to publish those attacks rather than to represent the complexity of the debate. Not much of any of that is helpful, either to humanity or to wildlife.

Anyway. The point is that conservation research can necessitate putting wild animals through the stress of being trapped and handled. The applications of these research findings form the backbone of our ethical justification for doing it. But while we think of ourselves as the good guys, I'm sure any given water vole doesn't see us that way. And if our investigations were just a matter of curiosity, or natural-historical interest, some of our methods, as comparatively harmless as they are, probably would not be appropriate. However well intentioned, even a live capture trap and the tiny implantable micro-transponders we use to tag voles have the potential to cause harm. The risks are worth taking because the survival and recovery of the species are at stake. The expected and hoped-for benefits can be weighed against the drawbacks. But with radio-tracking and water voles it is now clear the impacts are far too great, except in some exceptional, and very short-term, cases.

Our work has led to valuable behavioural-ecological insights. It has underpinned our knowledge of how to go about conserving water voles. But it has also handed us a salutary lesson in how complicated wildlife conservation can be – and how conservation scientists, above all people, must take the greatest care to do no harm.

5

Restoring
Ratty

The work of restoration cannot begin until
a problem is fully faced.

DAN ALLENDER,
The Wounded Heart

Put That Thing Back Where It
Came From Or So Help Me

MIKE WAZOWSKI,
Monsters, Inc.

I PULL THE CABLE TIE CLOSED. IT DRAWS THE WIRE MESH LID TIGHT, leaving just one corner invitingly open. I step back, then give our newly constructed release pen a clout with my boot. It responds with a solid *thunk*. Nice. I cast my eye down the line of pens. They look great: twenty plywood boxes marching off down the water's edge, 40 metres apart, open-bottomed, sunk into the soil.

'You know,' I say, 'I think these might actually work.'

Merryl, my partner in vole releasing, picks up her spade. 'Should do. I mean, they just have to keep some voles for a few days.'

Pragmatic as ever. But she's happy enough.

We carry our gear to where the components of our final pen are stacked on the riverbank. With our spades we score blade-deep troughs into the earth. Into these we slot the long edges of our plywood sheets. Then we give them a few hefty whacks each end with a rubber mallet to drive them down. Some good overhead clangs with the flat of a spade make it certain. With a quarter of the height of each sheet now sunk into the soil, they are gripped tight. The remaining, over-ground, plywood forms a rough, upright square. We drill holes in the corners where the sheets meet, insert cable ties and yank them taut. Then we stamp around, sealing every-thing in place before the wire mesh lid goes on. It fits well, with a bit more drilling and a lot more cable ties. We're done. Just like that we have finished our first-ever set of vole-proof – and predator-proof* – release pens.

'OK. So that was painless.' I look down at the day's worth of scratches the wire mesh has left on my hands. 'Mostly.'

'Don't be such a wuss,' says Merryl. 'You've had worse.'

'True.'

The buckets, spades and drills get slung in the back of the Land Rover. It looks a lot emptier than when we drove it down the field this morning, tipping piles of plywood on to an unsuspecting riverbank. And when we pull away it's with

---

* An important feature. We were not providing packed lunches for foxes.

the feeling that this has been a good day. We'll be back in a few weeks, once the pens have settled in. We have three more sites to prepare, but barring any calamitous interventions (in the form of, oh, I don't know, a colossal, steaming herd of someone's cattle trampling everything to bits),* this site is ready. The pens will be waiting.

We return, mob-handed. We jump from cars with our small horde of volunteers.† We set a team to writing numbers

---

* This happened at one of our sites. We discovered it a week before the release. A bad few days ensued. The landowner had failed to switch on the electric fence we'd set up to protect the pens. This left his dairy herd free to exercise their considerable curiosity, at length, all over our handiwork. I've long suspected that cows are actually quite intelligent and get dangerously bored hanging around a field all day. Their response to the 'entertainment' inherent in squashing our release pens confirmed this suspicion. The destruction was impressive. In retrospect it was probably the farmer's way of telling us that he'd changed his mind about being involved in the reintroduction, but in any case there was now no way we could let our voles go on his land. So we collected up the wreckage and left. In two frantic days, we had to find and negotiate a replacement release site and send a message to all our friends and colleagues begging for help. That weekend was spent digging pens with some of the most awesome conservation folks you could meet.

† Vole-unteers. (Sorry.)

on the pens, and another to ferrying bales of straw, plastic trays and lab cages to where they need to be. Conducting everything is Merryl, list in hand. Releasing a territorial animal into a confined space with some vole they don't know and won't like is really not an option, so back in the office we'd spent hours with spreadsheets and cups of tea to ensure it wouldn't happen. And now we just need to follow the instructions. Every lab cage is individually numbered, with stickers to tell us who is inside and which release pen they're meant to be in. On arrival the voles – bred in captivity for reintroduction – are tucked, in their cages, into the shady, long grass to the side of the pens. There they await their release.

But first the pens themselves need attention. In the interim weeks their plywood has warped, and reeds and sweet-grass have shoved up through the mesh. Some pens now gape at the seams. These need to be stitched together with cable ties and duct tape. Once they are vole-tight, we equip them with a chunk of straw, a tray of river water, carrots, apples and hamster food. It's enough shelter and provisions to keep our water voles for a week. But they shouldn't need anything like that long.

The whole set-up takes less than an hour. And when we're done we all gather, expectantly, by the first pen.

'Right,' I say. From a mixture of excitement and match-day nerves I'm grinning like an idiot. 'Shall we do this?'

Merryl and I grab the vole-handling gear and take a team of helpers each. We work down the line, leapfrogging one another as we go.* And after all those months of preparation it's a relief to discover that getting the voles into the pens is pretty simple.† Nerve-racking, yes, but simple. We scan their implanted transponders to double-check they are who, and where, we want them to be, then weigh and sex them before placing them into their pen. And that's it. There is just one (massive) pitfall to be avoided: 'hard' releases, in which a vole leaps gleefully from its cage and makes an ill-fated dash for

---

* Not literally. We had to at least *try* to look professional.

† You may well ask how we wouldn't know that it would be simple, given the amount of our lives we'd spent handling rodents. To paraphrase Helmuth von Moltke the Elder: 'No plan survives contact with the enemy, especially if that enemy is fluffy.' It's amazing how you can confect a foolproof study in the office then realize it's completely unworkable when you hit the field. You become permanently paranoid that you've missed something obvious but crucial that's going to ruin your life.

the river. Hard-released voles immediately find themselves lost in a brand-new environment and surrounded by hungry, sharp, toothy things. We built the pens to buy our voles precious time. The plywood walls are 25 centimetres deep in the ground, the mesh lid seals the top and the straw bale provides a nest. It's a temporary prison, designed to be escapable – with effort. To get out, in true British war-film fashion, the voles must dig a tunnel. That way, by the time they're exploring their new habitat, they already have a burrow and a refuge to run to.

Anyway, the fact was that after so many voles we could almost guarantee a faultless vole-handling demonstration. And yes, we, ahem, almost did. By which I mean that I completely stuffed up and one escaped. Water voles have a very low centre of gravity. Despite their legs being only a few centimetres long, they are somehow possessed of a manoeuvrability that fighter pilots would kill for. Once free, they are very, very hard to grab. Merryl and I once had to empty and dismantle the entire back of a Land Rover to retrieve a slippery rodent Houdini. And then there was the nasty moment when a vole sprinted across a car park and had to be corralled under someone's Toyota before we fetched it out. So when this particular water vole, halfway through being weighed, turned in her Pringles tube and started heading for the light, I knew we were in trouble. I grabbed the tube off the scales and clamped a glove over the end. But she thrust between my fingers, grasped the tube's open edge and hauled. She squirmed out and perched for an instant, all four feet on the rim. I grabbed for her tail just a millisecond too late. She leapt and landed on the grass at our feet. In the moment it took me to recalibrate, she had legged it. She disappeared straight into a giant stand of stinging nettles.

My first thought was, 'Oh, she's never coming back, then.' I didn't have time for a second thought before one of my helpers dived – literally dived – headlong into the nettles after her. A brief scuffle and he was up, holding an indignant vole by her scruff. I gawped. Then I dumbly held the pen's lid open so he could put her inside.

'How did you do that?'

The guy, a colleague from another research group in the Zoology department, shrugged. 'I've just been in West Africa catching snakes.' He smiled. 'Venomous snakes.'

'Ah.' I nodded. 'Yeah, that would do it. Thank you.'

That vole was the exception. Everything else went beautifully. Really beautifully. At each pen, when we were done with the measurements, we passed the tube and gloves to a volunteer. And they took the tube and looked delighted. I mean, in most contexts that would be quite a weird thing to be handed – essentially a rat in a snack packet – and potentially the cause of some distress for the recipient. But on the riverbank everyone held the Pringles tube as though cradling something incredibly delicate and precious. They eased it, carefully, through the open corner of the mesh, while everyone else clustered round. Sometimes the vole would sneak into the watching hush, then rise up, peer about and clean its whiskers. Sometimes the exit was a slide and bump, and a bemused waddle towards the shelter of the straw. Often the vole panicked and jumped, before scuttling around the rim of the pen. But every release, every release day, was greeted with relief and smiles.

When we were done, we closed the pens fast shut. At the end of the afternoon we collected up the holding cages and tools and carried everything back to the cars amid laughter and chattering. We felt wonderful about it. We had put creatures back where they belonged. We had given them a chance to be wild. And even a decade later, that makes me happy. I can understand folks who buy captive birds just for the joy of setting them free.* The action feels like righting a wrong. But the joy of the release can never be unalloyed, because you can never be certain of the outcome. You could see that, too, in the way our assistants sometimes glanced uncertainly back at the pens. Their look said, 'Brilliant, we did it! So what happens now?'

The answer was that I really didn't know, and neither did Merryl. I knew we'd be back to check on things in a week's time. And the next month we would set our traps, to monitor,

---

* I would never do it, though. People being what they are, the markets for captive songbirds – often sourced from the wild and ready for purchase and release at the customer's convenience – have caused the wild populations of many species to plummet. Most of the birds end up being immediately recaptured for the next punter, anyway. There's no good intention that won't be hijacked by someone else if there's a profit in it. And also it's worth saying that unless you have engaged in quite a lot of preparation, a typical outcome of setting captive animals free is that they will blunder about until something eats them or they die of starvation. The welfare consequences of 'free but unprepared' can be horrific. Most of them probably won't make it, and so you're best not getting involved unless you're working with professionals.

weigh and measure the voles at our release sites. But what would we find? Would it be banks thronging with furry life, or empty but for some mouldy straw and blackening plywood? This was an experimental release, meaning the whole project was a colossal gamble. That's why I had already spent four years of a Ph.D. studying water voles before attempting one, and why it was the basis of Merryl's doctorate, even after she had spent a decade working on small mammals. As a project it was neither simple nor risk-free. We hoped, and had good reason to think, that it would turn out all right. But there were no guarantees.

Our plan was to reintroduce water voles to twelve rivers in Oxfordshire. In so doing we would reverse their decline in the region for the very first time. That on its own would be hard enough, but being part of an academic institution meant that we had to go further. Our boss likes to say that it's hard to be interesting, and even more difficult to be useful. We used to wish he wouldn't – mainly, but not exclusively, because it often meant a lot more work. In the case of our releases that sentiment meant we had to set everything up so that our results could provide tests and data from which we could draw clear conclusions. We had to run a giant reintroduction experiment so other people, even those working on other species, could learn from our study. We had to examine fundamental assumptions about how habitat quality might influence the chances of success in reintroductions – not only of water voles in Britain, but reintroductions in general, any-where in the world.

And so our release sites couldn't just be random lengths of river. Instead each was a self-contained 800-metre length of water vole habitat. This was the minimum my statistical models had told me could support a vole population in its own right, but we made sure that each site had more good habitat within easy vole-dispersal distance. We selected sites based on how wide the vegetated margins either side of the river were. At six of our twelve sites we chose narrow margins. The vole-friendly plants stretched anywhere from 1 to 3 metres away from the water. At the other six sites there were lush margins a full 3 to 6 metres wide. These widths are pretty typical of the ranges of vole habitats across lowland Britain, and selecting within these categories ensured that our 800-metre lengths of river were a good representation of the types of habitat quality available to water voles nationally.

But before we could even consider releasing voles into our

sites we needed to make sure that we had removed the reason they weren't there in the first place. You wouldn't believe the number of reintroductions where people haven't bothered to deal with the original problem. They set their species loose, hoping to repopulate an area from which it was extirpated. And then they watch in horror as exactly the same thing happens all over again for exactly the same reasons. I mean, who'd have thought it, eh? To avoid this in our case meant above all keeping American mink well away from our voles. We already had a local mink-control programme, set up and run by one of my colleagues and the local wildlife trust. Working with them, we extended the control to 4 kilometres upstream and downstream of our release sites. And then, over three years, 2005 to 2007, into our carefully positioned sets of release pens, we let our water voles go.

The results were amazing. Our released voles were born of captive parents, in nest boxes in cages with mesh floors. They had always been fed daily. They had never needed to forage. Before the release they had spent weeks in plastic lab cages. But somehow, when their paws met soil, these tiny adventurers knew what to do. They dug. In short order burrow entrances popped up on the other side of the pen walls. And in other places strong teeth attacked pen corners, chewing through plywood rather than earth.* However they escaped, the result was that after just a few days nearly all of our voles were out and about. And when they found themselves next to a river, something inside them kicked into action. They ate plants and deposited feeding sign. They created territories, and left latrines to mark them. They made runs through the plants. They dug (chewed) extensive burrows. They foraged and survived. When, after the first month, we set traps to see how many voles made it, we were able to check who had survived, and their weight and health. By the second month we were already catching brand-new juveniles. Truly wild after just one generation, these youngsters had grown up on the

---

* Fun fact. You might assume that water voles use their paws to dig. This assumption would be entirely reasonable, but wrong. In fact, they use their teeth to chew through the earth. They avoid getting a mouthful of soil by having a diastema – a large gap between the front incisors and the rest of the teeth – into which their cheeks can be pulled to create a barrier. Regardless of the impressive dental arrangements, it's a very weird thought, when looking at how extensive water voles' burrows can be, that all those tunnels had to be chewed out of the ground. This is just one reason among many that I don't want to be reincarnated as a water vole.

river. Overall, as a result of our efforts, seven re-established populations spread down through the river habitats, colonizing as they went.

And we learnt a lot. We discovered that water voles released into the sites with the broadest swathes of vegetation stayed closer to their release pen, and were more likely to survive than those released into narrower habitats. We ended up with good initial post-release monitoring data from nine of our twelve releases (more on the other three in a second). Into these sites we released a total of 396 water voles. Just under 50 per cent of them survived the first month. Normally we would expect about 60–70 per cent of a natural water vole population, regardless of their sex, to survive that period. It seemed that the release process had taken a toll. But when we looked more closely, we saw that actually 60 per cent of our females had survived. It was just that only 40 per cent of the males made it. Which left the question of what on earth had happened to the males that hadn't happened to the females?

The clue was in the distance they moved away from their release pen. Males on average were caught about 100 metres from where we let them go, while the females were only 50 metres away. These distances were also affected by the width of vegetation at the site. The narrower the margins, the further voles of both sexes strayed. The further they strayed, the lower the chance of them surviving to be caught in our traps. At the sites with the narrowest margins, only about 45–50 per cent were recaptured. At the sites with the lushest and broadest swathes of vegetation, we got 70–80 per cent of them back.* But after the first month, all sites showed 'normal' survival rates for wild voles. So while we can't conclusively prove the cause-and-effect relationship, the coincidence is far too suspicious to ignore. It looks as though a released vole faced with a serviceable, but underwhelming, future home will take a hike to find something better – and while doing so is more likely to be nabbed by a predator. Any voles that pop out of their release pen into a thick, luxurious fringe of green will stay put, and prosper as a result. And because males tend to run around more anyway, they will be relatively more at risk in any habitat until they get settled in.

---

* It was possible that the voles at the narrow sites dispersed out of the trapped area, and so appeared to be dead when they had actually just left. But only a very small percentage of any population disperses like that. Sadly, in most cases, the voles we didn't recapture had probably died.

As the months progressed we recorded how vegetation abundance affected the size of the populations that each site could support. Those with the least vegetation had two or three adults per 100 metres. Those with the most had four or five. It might not sound like a big difference, but throw in a load of babies and juveniles, and repeat this pattern over kilometres of river, and it adds up to a far, far larger population. And so our experiment gave us concrete evidence supporting the long-standing rule that any reintroduced animals need to be given the very best habitats we can possibly provide. There is a difference between surviving and flourishing. Water voles do more of both when given just a few extra metres' width of plants to live in.

To our massive relief the releases were a conservation and research success. But I'd be giving you entirely the wrong impression if I made out that everything we attempted went brilliantly. Because it didn't. Advancing knowledge in a subject means doing stuff that nobody else has tried.* And this

---

* Or, to put it the way most researchers do, 'This could be cool – do you think it'll work?' and 'What's the worst that could happen?' The honest answer to both of these questions is usually, 'I don't know,' often followed by, 'But what the hell, let's do it anyway.'

turns most research projects into a potentially career-ending bet on the outcome. If enough goes wrong you could end up with literally nothing to show for three years of funding – and no career, because what funder wants to see nothing in return for their money? So you quickly learn to spread-bet. You can build in lots of different elements to one project, in the hope that at least some will turn out well. But nearly everyone has experienced the creeping horror of discovering that some portion of their carefully collated results has mysteriously turned to trash while they weren't looking. And yes, this happened to us.

For me the dawning comprehension came, appropriately, while I was standing in a meadow and gazing up at a changeable sky. It was one of those British mornings that starts out a perfect, brittle blue but throughout which clouds bubble and gather before the sun eventually goes in. I watched the clouds swirl together. They fused, snuffing out the sunshine. And I realized we were probably in trouble. I glanced back down, and suspicion became a gut-punch.

'Merryl?'

She didn't reply. She was seated in front of the open rear door of the Land Rover, bent over an anaesthetized water vole. She was measuring foot length, body fat and urine concentration, and taking samples of faeces and a tiny drop of blood. By the time the sun disappeared we were already halfway through a morning's trapping. We were using a novel technique in a way that had never before been attempted, one that had cost us a colossal amount of work in the preceding months. But the results had been brilliant. They alone justified the succession of weeks of twelve-hour days of exhausted trudging, labouring and bickering. They alone were our smug pay-off for having bolted hours of additional sampling and measuring on to already stupidly demanding field days. But now I was convinced that these results, the basis of our continued sanity and happiness, were, in fact, an utter load of old cobblers.

'Merryl?'

'Hang on, let me finish this vole.'

'Merryl, we have a problem.'

'How serious?'

'Well, nobody's going to die. I hope. But, um . . . yeah, not sure. We need to talk.'

Merryl had kitted out the Land Rover as a portable laboratory. Thanks to her efforts we could drive to site, collect water voles from our traps, quickly anaesthetize them and take an assortment of measurements – before bringing them round,* giving them an apologetic chunk of apple and letting them scamper free, right back where we caught them. The laboratory had everything. It had electric power, provided by a stack of interconnected batteries, a fridge, a gazebo (intended to keep us at least nominally dry), a full anaesthesia kit (vole-sized, comprising a bottle of liquid anaesthetic, a cylinder of oxygen and a regulator to mix these), a luminometer, a bio-hazard bag, a sharps bin, and assorted tubes and bottles. And one of these pieces of equipment was responsible for ruining about two hundred man-hours of research effort. Any guesses which?

Yes, it was the traitorous luminometer. Seriously, that damn thing had one job.

---

* I should say that we and the project were in possession of a *lot* of different licences to be permitted to do this. And the anaesthetic was amazing stuff. It's gently administered as a gas by placing the whole trap in a box. The vole is out cold by the time you open the trap. When you want to bring the vole round you just take the gas away. It's alert and happy in a couple of minutes. I accidentally once got a lungful. It made me a) a bit giggly and b) feel like I should have a little sit-down for thirty seconds or so.

It is not easy to explain how a luminometer can ruin your year. It's not even that easy to explain what a luminometer does, exactly.* But as a start, have you ever had two hard weeks at work, struggling to meet an impossible deadline while everyone is piling in more requests? And then somehow, when you finally manage to get everything done, printed and handed to your boss, you wake the next morning feeling like you've been run over, with the sniffles and a sore throat? If so, then you've experienced exactly what we were trying to use it to measure: the effect of chronic stress on the immune system. Just in our case it was with water voles.

It's comparatively simple to tell if humans are stressed. Sometimes they even tell you, at great length when you'd prefer them not to. With animals it's harder. Studying their behaviour works fine when they are in captivity, but not so well with small, cryptic beasties in the wild. And often if you're getting close enough to watch an animal's wild behaviour then the thing stressing them is you. So how on earth were we

---

* From the online medical dictionary: 'A luminescence photometer used to assay chemluminescent and bioluminescent reactions'. So there you go. Clearly something with unlimited potential to wreak havoc.

supposed to tell if our water voles were chronically stressed? (I'll answer the question of why we'd want to know that in a second.) We opted to measure the state of their immune system because – to cut a convoluted story very short – being stressed can stimulate your white blood cells (leukocytes) in the same way as if you are combating an infection. And that can wear them out. A few weeks of high stress can deplete your reserves and seriously lower your ability to mount an immune response. We can measure this by taking a drop of blood and making it think it's under attack by bacteria, and mixing it with a chemical that luminesces when the white blood cells are active. We measure the tiny amount of light emitted using a luminometer.* In general the less stressed your vole, the better its leukocytes will function and so the higher the reading of Leucocyte Coping Capacity (LCC).

We had previously used LCC to study how water voles coped with being captive. We found that voles kept in outdoor enclosures, with water to swim in,† were happier than voles

---

* Or, to give it its full title, an 'unspeakable wretch of a luminometer'.
† They mainly used the water as a toilet. But as long as they enjoyed themselves I feel that what they did with it was, quite literally, their business.

that had been put into laboratory cages. The lab-cage voles progressively lost weight, became more dehydrated and gave lower LCC scores than the enclosure voles. We also attached dummy radio-collars to twenty of them and found that their weights and LCC scores dropped in the first week, compared with uncollared voles. So while lab cages are essential for moving water voles to a release site, if we want our voles happy and fat then we should keep their time in the cages to a minimum. Also, it confirmed our previous explanation about the radio-collars and sex ratios: they really did stress the voles.

The success of this mini study gave us an idea. Wouldn't it be great, we thought, if we could measure the LCC of water voles before we let them go, and then get the same measures post-release from the field? We could then prove something that many people have suspected for a long time, but for which nobody really has any solid evidence: that animals in their natural habitats are far happier than in captivity. So we measured our voles' LCC in lab cages, days before the release, and the results backed up what we had previously found. We got higher scores from voles that were in a cage on their own, but progressively lower scores from those housed with two or three or four (or even eight) siblings. The more space each vole had, the happier they were. Not a surprise, but again, good to have proved.

The fun really started when we began collecting field LCC measurements. Merryl took the samples and I ran the luminometer. I hate lab work.* And now I had to do it literally in a field, armed with micro-pipette, cuvettes (square, clear plastic vials that go into the luminometer) and batches of liquids in light-proof tubes. Each time Merryl took a blood drop I had to add it to the cuvette, get the reaction going, put it into the luminometer, close the forever-cursed-thing's stupid-damned-light-proof lid and take a reading . . . all of which were off the chart. They were double, triple what we had been getting when the voles were in cages.

This was a stunning, massive vindication of our dedication. We thought we had just shown that newly wild voles had become far more immunocompetent. Encouraged, we kept

---

* Because I'm clumsy. For one early project I was put in charge of a vial of liquid that came with the following health warning: 'Danger, may cause heritable genetic damage. Avoid all contact with skin.' Yes, this stuff was so nasty that just one splash could affect your unborn children. So of course I spilt it all over the bench. I narrowly missed my labmate. Do not put me in a lab.

going for weeks. But some days we got weirdly low readings, while on other, sunnier days, the readings went high. And finally, on what became the last day of our LCC measurements, the clouds covered the sun at the precise moment when I was about to take a repeat measurement on a sample. The same blood from the same vole gave a substantially lower reading than it had 30 seconds before.

Oh dear.

The weather closed. We were spritzed with drizzle then briefly lashed by sheets of rain that spattered on to us from around the edges of our gazebo. The next LCC readings were all low. And when the clouds bundled away and the sun sprang out, the readings went high. Very high. We placed the luminometer under a cardboard box and retested. The measurement went low again.

And that was that. The 'light-proof lid' of the luminometer wasn't. We had invested two hundred man-hours and hundreds of pounds into a detailed set of results that demonstrated that a sunny day is brighter than a cloudy day, and both are brighter than the Zoology department. It is discoveries of this calibre for which the word 'poopants' was invented.

We mentally discarded months of work. We finished the

day's trapping, went home and opened a bottle of wine.* The only good news was that our two published LCC studies had been conducted under stable indoor lighting conditions, so they were trustworthy. And now, years down the line, the whole saga mostly doesn't register. It was a risk we accepted by trying to do something novel. Which isn't to say that I have nothing to regret from our releases. I am proud of our reintroduction work, but still feel terrible about the voles that didn't make it. Nine of our twelve populations initially established on site. Eight of these bred. Seven survived the winter, and expanded the next year. But that means we lost five. Of these, one was hit by freak flooding in the first month. The river rose up and wiped it out. It was so fierce that it carried away the release pens. The whole region was hit and there was nothing we could do. (Although it's some comfort to know that there are voles living there now, spread to the site from another of our releases.) But the rest . . . well, despite all our careful arrangements, the mink got them.

---

* Not only did we work together but at the time we were also housemates. Somehow, despite our 24/7 existence, Merryl didn't murder me. I'm still grateful.

A hard truth about reintroductions is that they are new and uplifting and can galvanize people to take conservation actions – in the short term. We got a lot of help setting up our mink-control areas. Volunteers and landowners signed up to help check mink rafts, set traps and remove mink. With their help we were able to clear areas into which we could release water voles. But by the time we got to the releases people had already been checking rafts for months. Some might not even have seen a mink in all that time. And so, as the year progressed, many folks found the routine of checking mink rafts and traps burdensome, especially if they weren't finding anything. Other priorities began to take precedence. Weeks began to elapse between checks for mink footprints. In those gaps between diligence and dispersal, suddenly some of our voles were in big trouble. And, to be clear, I absolutely do not blame any of the landowners or volunteers. They were doing us a favour. No, the buck stops squarely with me.

It does not feel good to walk down an unoccupied riverbank, uprooting release pens with a spade, knowing that you could have avoided this. I collected plywood sheets that were twisted and good for nothing, and mesh that was hung about with snipped-off cable ties. And I knew that one glance at the riverbank would reveal the lingering signs of the voles' brief wild existence. Their memory was there in the gnawed ply-wood and the grassed-over exit holes. It was in the rotting, yellowed feeding sign still heaped in disintegrating piles; or perhaps in the rain-pocked remnants of a dried-up latrine. Some signs still looked fresh. They spoke of voles that until two weeks past had been hanging on in enclaves. They spoke of a duty of care poorly executed. For one brief instant in their lives these animals were out and free. And then they were prey. It's never fun to know that you have let somebody down, even if that somebody is a water vole. In fact, in some ways it's worse. The water voles have no share in the respon-sibility, and no comeback. They accuse you only with their absence.

Most of my memories of my time in the field aren't like this. Most are wonderful.* Only a handful are absolutely hor-rible. The very best and worst of all these involve mustelids, the family of carnivores which in the UK contains stoats,

---

* OK, so actually most are of being irritated by rain, bites, cuts, grazes, horseflies, wasps and mosquitoes. And luminometers. But a lot are wonderful.

weasels, polecats, pine-martens, otters, badgers* . . . and, of course, American mink.

The happiest thing I think I have ever seen was a stoat hunting along a dyke edge. Wildlife researchers shouldn't (but do) anthropomorphize, but surely nobody could witness how completely and joyfully absorbed a stoat is in the moment – bouncing, pausing, scenting the ground before bounding ahead, snuffling and racing to the water's edge then leaping in, half paddling, half slinking across the surface, jumping out, shaking off, then sprinting away – and not find themselves caught up in the sheer jubilant thrill of its focus. You cannot help but feel some echo of its pleasure.† Everything about stoats tells you they just love to hunt.

My angriest wildlife sighting was a weasel that had blundered into one of my water vole traps. It probably went in

---

* Most mustelids, apart from the badgers, are quick and agile and actively hunt their prey. Badgers, for various reasons, often elect to eat worms. This is a bit weird, and the underlying reasons for the, um, diet of worms have taken researchers years to understand. But all mustelids share one common trait: they are, under no circumstances, to be messed with.

† I found myself accompanying its movements with a mental *Doink, doink, adoinkadoink doink adoink*. But that might just be me.

expecting there to be a meal at the end of the scent it was following, but instead spent a hungry couple of hours unable to
get out. It used that time to get properly cross. When I opened
the door – from a safe distance, with a stick – it dashed a few
paces out and stopped dead. It raised its head and glowered at
the world. It vibrated with indignation and – I'm not kidding –
stamped all of its feet in fury. Then it screamed the weasel
equivalent of *Aaaarrrggghh!* and ran headlong, tiny paws
blurring, into the undergrowth. Weasels are smaller than
stoats, but they compensate.

And I cannot forget the first time I saw a mink.

I walk towards the mink trap. Inside is the mink, which
hisses at me, and runs up and down the cage. I hesitate, swal-
low, then approach. I don't want this, I think. I'm the one who
set the trap, but I don't want this. One of my hands is thickly
gloved. It holds a pair of wooden combs, with which I gently
crush the mink against one end of the trap. Cold in the other
hand is an air rifle. I don't know if I can bring myself to use
it. I lift the barrel, hold it point-blank to the mink's head. And
I can't. Surely I can't. But then the oddest thing: the mink
stops struggling. It lowers its head, almost in acknowledge-
ment. My finger tightens.

The rifle does its job. I shoot the mink twice more, to be sure it isn't hurting. It twitches the first time, not the second. I drop the gun. I stand. I slump with my back against a tree. I spend the next five minutes crying. Then I get up and reset the trap. It now smells of mink, and will be irresistible to others nearby.

The next day I catch another mink. I shoot that one, too.

It was necessary. It was. That's what I have to tell myself. The mink had to go, so that we could restore the voles, and learn what we needed to learn to save them. And perhaps I could make myself feel better by pretending that mink are villains, needing to be despatched. But that's not true. The American mink are not villains. They are dual victims of human carelessness – of farming them in a country where they never belonged, and of then letting them get out – and of their own spectacular evolutionary success. They are fantastically adaptable. They can hunt and eat a variety of animals, from rabbits to voles, to fish, to birds and their eggs. They can disperse for tens of kilometres, and so spread with ease to new territories. They can alter their behavioural patterns from being nocturnal to diurnal to escape aggression from otters.* A female mink is tiny compared with a male, which is probably useful in preventing the sexes from competing for food (she can slip down a water vole's burrow where a male cannot) – but also means that a female mink is unable to be selective in which males she mates with. Her solution is to mate with all of them but to delay implantation of the fertilized embryos until she has a good collection from a number of candidate fathers, and then, somehow, to internally select which ones she would like to see as offspring. Her litter of kits can have individuals from multiple fathers, a great tactic for ensuring the best genes are put out there to face an uncertain world. Ecologically, they are magnificent.

---

* Yes, about the whole otter/mink thing. Otters have returned to prominence in British rivers. And the conservation community should absolutely own up to having entertained the cowardly hope that their newly restored populations would do us a favour, bump off the mink and save us the ethical burden and financial cost of mink control. Early studies did show a decline in mink signs in newly otter-recolonized rivers, and we grabbed this result with both hands. It took years more for us to realize that the mink were actually alive and well, just keeping a sensibly low profile. In the absence of human intervention the mink are, I'm afraid, here to stay.

Mink are feisty and lightning quick.* They are brilliant and lovely. And that is a terrible thing, because it would be far simpler if they weren't. After all, people become conservation scientists because they love wild things. We want to save them. But conservation contains few easy choices. I wanted to reintroduce water voles. That meant I had to remove the mink. And there was nobody else who could check this trap for me. I was left to do my own dirty work. As a result I sacrificed something exquisite, and also some part of how I perceived myself. I'm the one who loves wild animals so much that I want to save them. But here I was, gun in hand. It is a representation in miniature of some of

---

* Merryl once discovered this to her cost. One moment she was handling a mink on a riverbank, the next it snapped needle-fine teeth into her thumb. The latex glove she was wearing filled with blood, which began to drip out from the puncture holes. At that point, apparently, she began to feel a bit light in the head. We've both been munched by a lot of wildlife, mostly rodents, and usually it's not so bad. It bleeds, but you can clean the wound with some alcohol and expect to be fine. A mink bite isn't like that. That evening Merryl went to hospital, with red infection lines tracking up her arm's lymphatic system. One of the drawbacks of being bitten by a carnivore is their dubious oral hygiene.

the choices we face nationally and globally about enacting conservation. Years of accumulated water vole research point to the same conclusion: if we want water voles in Britain, the mink, as beautiful and blameless as they individually are, must go. That's an absolute prerequisite. We can't have it both ways.

And this is a problem. It means that for all our reintroduction successes, we have been left with more questions. Will those of our reintroduced vole populations that survived to breed and spread still be here in ten years? Or will they too, one day, succumb to exactly the same mink-related factors that claimed their predecessors? I said earlier that you would not believe how many reintroductions don't remove the original cause of the decline. We put in place every attempt to do just that, but still it wasn't enough. Because despite ongoing mink control in Oxfordshire, our populations remain vulnerable to the slightest lapse. There is a giant, national population of mink out there, just waiting to flood in.

And so this, finally, is where all our water vole research brings us: right back to American mink. To plan any future for British water voles means engaging with the future of American mink. The fact is that as long as we have not removed mink, we cannot have saved water voles. The existence of any vole population remains contingent on the absence of their fiercest predator. It doesn't matter if the population is newly released or has been there for thousands of years, we have failed, and continue to fail, to deal with the underlying conservation problem. Our water voles are still threatened, every last one of them.

We don't have a clear figure on how much it could cost to eradicate mink from Britain. My own back-of-the-envelope calculation was that it could cost £60 million and take ten years. But my estimate may well be low by many tens of millions. Still, the point is that mink removal is eminently doable. Thanks to amazing projects that have researched and conducted mink removal we know we can make it happen. And then with a few targeted reintroductions – which again, thanks to a lot of collective work, some of it my own, we know how to do – water voles would return quickly to repopulate much of the country. And we'd never have to spend another penny on them. It's an easy conservation win.

But what government on earth will stump up £60 million, or more, to save some rodents – even if those rodents are one of the most loved fictional heroes of all time? Not ours, that's for sure. And in the interests of full disclosure I'm pretty certain that we, the conservation community, have never presented our government with a bill for the actions needed to end the water vole's conservation problems. But this is largely because, from bitter experience with a whole host of species, we know that it's pointless. That money is not a conservation priority. And if we had a spare £60 million for British conservation we could almost certainly create a far larger impact spending it elsewhere, on something even more desperately needed.

And so we find ourselves at the end of our conservation research journey, solution in hand. For water voles we long ago reached the stage in our idealized conservation project trajectory where the implementation and management should start. I mean, this was supposed to be the good bit, the bit where we roll out the cure on a national basis. I can so easily see how it would work. We would start small, then gain momentum, learning and monitoring and refining as we went. And eventually, perhaps even after five to ten years, we would have done it. Most mink would already be gone, and the otters would probably take care of the stragglers. But just to be sure we would put management in place, at key locations, to monitor things and keep our voles saved. And soon it would be job done.

None of that has happened, of course. Instead the status quo continues: patchy money, annually drip-fed, keeps the mink away from our few surviving and most precious populations. In some places we are winning. Swathes of the Highlands are mink-free thanks to a brilliant project up there. And there is mink control across much of the south-west of England, and a new project being scoped in the East Anglia fens. But all are reliant on funding. And funding, in conservation, is never guaranteed. All it would take is a hiatus, even of a year or so, and the mink would patter back down the rivers and begin anew the process of masticating voles. We could quickly and easily find ourselves in the weird position of possessing a clear and workable solution – for one of the first species for which the ideal conservation trajectory has been followed right the way through to implementation – but far worse off than ever before.

What does all this say about the practical worth of our research? Nothing good, I fear. At my parting of the ways with water voles this was the outcome of those decades of research, the work, the discoveries, the populations re-established, the optimism. 'Ratty' remains far from restored. He's on life support, fed by a drip.

6

River Red
in Claw

Throughout history, men have tried to play God by moving rabbits, goats, sparrows, mongooses, and a hundred other species to oceanic islands and island continents, and later have wished to God they hadn't.

VICTOR BLANCHARD SCHEFFER,
The Year of the Seal

Sandy beaches still rim the lakes, but if Lake Michigan, for example, were drained it would now be possible to walk almost the entire 100 miles between Wisconsin and Michigan on a bed of trillions upon trillions of filter-feeding quagga mussels.

DAN EGAN,
The Death and Life of the Great Lakes

THE BLUE NYLON ROPE DROOPS UNINVITINGLY DOWN THROUGH THE water's surface. The river here is still, and turbid. I dump the rucksack and builder's tub I am carrying by the peg that secures the rope to the banktop. Then I bend down, take the rope and start hauling.

I pull slowly. Mainly because I know what's down there.

The encounter I'm delaying is an echo of one from my early days of water vole work. I had been between traps when something caught my eye: a crimson flash, glimpsed through the grasses. There was something vivid near the water's edge. So I padded a little closer. Yes. There. Lurid scarlet, and now a suggestion of turquoise. I brushed the grasses aside and knelt beside a collection of hard objects, like broken bits of a child's toy. They were brown, blue and crimson and loosely piled on the bare earth of a water vole's run. Weird. The things had been left just where you'd expect to see water vole feeding sign. But they were nothing like the usual chewed-up lengths of plant stem. I gingerly pick up a piece. It was a chitinous claw, hinged like a lobster's and the size of my thumb. Inside it was hollow. Something had taken the meat out of it. I had never seen something like this before but knew what it must be. But how had bits of dismantled crayfish got this high up the bank? Maybe they had been left by a bird or something. But no, I spotted a water vole latrine, and a little beyond it more tell-tale shards of crayfish carapace, this time unmistakably mixed with vole-chopped plants. So. My furry little not-so-herbivores really had been out crayfishing. Score one for the mammals, I decided.* And I thought no more of it.

Until now. The rope passes steadily through my hands. And soon I spy a shadow rising up towards me through the murk. I see the vague outline of a fat, dark cylinder . . . which

---

* Water voles are the original flexitarians in the sense that they are herbivorous 99.9 per cent of the time but are never going to pass up free protein. They will happily grab a signal crayfish, given any opportunity. Similarly I have also found a pair of gnawed frog's legs left among feeding sign. I sincerely doubt the vole caught a frog, but it may have found one lying around dead and thought, 'Well, it'd be a shame to waste *that*.'

resolves into a two-foot-long, black plastic mesh tube with yellow funnels, topped with a jaunty, white licence tag. Then it breaks the surface and goes instantly heavy. The sheer, dripping weight of the thing tells me I'm in for an interesting few minutes. I lower it down to the bank. For ages I do nothing but watch the movements inside. Brownish, rounded, glistening, clattering forms shuffle and snap at one another, jerkily banging against the mesh. My crayfish trap did its job well.

Keeping my fingers clear of the holes, I unclip a funnel. Then I upend the contents into my tub. Crayfish tumble out, ten or twelve bodies, all big, each the size of one of my hands. They *thunk* on to plastic, wriggle for a bit and right themselves. Some immediately tuck in their legs and go catatonic. Others stagger about or tussle, grabbing at each other with outsized claws. I watch them with fascinated dismay. The very last thing I want to do is put my hand in this tub. But then what's the point of being an ecologist if I'm too scared to handle my study animal? So I take a breath, and reach in.

Crayfish rear up. Claws wave and flex. I whisk my hand away. Then I feel like an idiot. Come on, Tom. You're hundreds of times their size. I tip the bucket, sending most of the mass skidding into a pile, and grab for a straggler. I dart behind its claws, pinch it firmly by the thorax and hoist it aloft. And just like that, between my fingers and rough to the touch, is my first-ever American signal crayfish, Latin name: *Pacifastacus leniusculus*. I admire its brown, spiky carapace. I turn it over to examine the whitish underbelly, all body segments, legs and writhing swimmerets. From below, the hinges of the claws have patches that are a startling turquoise blue. The claws themselves are blood red. I look the thing in its black, beady eyes, then chuck it back into the water. I take a breath, and turn to the others . . .

My friend Rosie and I disagree. She thinks signal crayfish are 'pretty'. I think they are vile. Wonderful, but . . . vile. En masse they look like something from *Aliens*, or *The Dark Crystal*. Your first reaction is that they shouldn't be down there.* Your first reaction is correct. They shouldn't. They are yet another non-native species out and about in the British countryside and causing problems for the locals. My job, once more, was to see if we could sort out those problems. So

---

* Most people's second reaction is that if they have to be down there, they should *stay* down there.

despite my reluctance, getting that first trap done was a relief. It meant my new study had legs. Lots of legs.

In early autumn 2008 I officially became a carcinologist, a researcher of crustaceans. I'll admit this wasn't the career path I'd imagined at the outset of my studies. I had grand visions of completing my work on water voles, then moving on to other mammals, possibly (whisper it) in exotic locations. But the realities of research, especially in conservation, can be stark. Contracts are fixed term and short, and funding hard-won. You get three years if you're lucky,* and heaven help you if you haven't lined up the next money by the end of

---

* Which in academic circles is akin to lucking out on a lottery scratch card. For everyone with a Ph.D. who is doing science in Britain, 3.5 per cent will end up with one of the breathtakingly rare permanent positions. Everyone else scrapes by on short-term, often annual, contracts. We bring in money, which we give to our university. The university then uses this to employ us. If we fail to bring in money for a given year, our contract ends and we have to leave. Which costs the institution very little, given that we are easily replaced by someone cheaper, newly qualified and who hasn't learnt that universities are essentially running a pyramid scheme. Academia is fun while it lasts, but for most of us it's simply too risky to be a vocation for life. By the time you're in your forties, the clock is well and truly ticking.

it. Which is another way of saying that a researcher's career path is determined not only by the urgency of the research (there are a lot of urgent topics out there), but also by which questions will suit your skills and pay your salary. In my case my vole project ended just as we secured three years to study British crayfish. The puzzle was intriguing, the research vital, and with my riparian field experience I was a natural fit. So my job became finding a way to save our native species, the white-clawed crayfish.* Signal crayfish had been spreading for decades through our rivers. In the process, they had obliterated population after population of white-claweds. We needed to know if this was as bad as it looked, or could white-clawed and signal crayfish coexist, if the conditions were right?

I had a simple plan. The Environment Agency had good, but patchy, distribution data for both the white-claweds and signals in Oxfordshire rivers. The data showed populations of white-claweds holding out in upstream locations, while the

---

* Their Latin binomial is *Austropotamobius pallipes*. Taxonomists will never come up with something short and memorable when there's the slightest possibility of a polysyllabic nightmare.

signals encroached from downstream. My job was to fill in the gaps. The data collected would let me compare the presence and absence of both species with a variety of habitat measurements, like the depth, flow rate and turbidity of the water, the amount of oxygen it contained, the amount of tree cover, and the types and abundance of other aquatic macro-invertebrates (all the largish aquatic invertebrates – like freshwater snails or fly larvae – that you would expect to find contributing to a healthy river). We hoped to discover some combination of habitat characteristics that benefited the white-claweds enough, or were sufficiently detrimental to the signals, to let the species coexist. Then we could identify similar places across the country where white-clawed crayfish could be protected.

I prepared diligently. For initial training, I headed up to the Lake District. There I spent a lovely day learning to catch white-clawed crayfish. The river in question was shallow, its flow bumping and tumbling over cobbles and rocks. Stepping carefully, in my (scrupulously clean and borrowed) waders, I am shown how to position a net downstream of a likely looking rock, then bend down and grasp it while gazing through the surface shimmer to focus on the objects below. Then, taking the utmost care not to crush any creature hiding beneath, I gently tip up the rock. Nothing happens, so I pull it loose. A few bits of grit detach and are whisked away. Under the rock is just an empty cavity. So I replace it, shuffle a few steps to the side and grab another, in a slightly deeper pool. Net ready, I hoist the rock up. A white-brown form, not much bigger than my thumb, scoots up, up and into the flow of the river. I drop the rock in surprise and fumble for my net. But too late. The crayfish has already sped beyond my reach. It performs a perfect, graceful arc, plunging down towards a new and distant hidey-hole. It vanishes into the shadow of an unseen boulder and I grin sheepishly at my instructor.

'Amazing, isn't it?' she says. 'And to think they do that backwards.'

For a moment I replay what I just saw. And yes, the crayfish had been propelling itself backwards through the water, powered by its tail: a quick snap to curled, then release, snap and release. And all with the big front claws left trailing and streamlined for minimum resistance. The action I had perceived as a smooth arc had in fact been a series of rapid tail jerks, each yielding a change of direction, homing in on its chosen destination. A diminutive crayfish reverse parking, at speed, under a rock in a fast-flowing river. That's quite a feat. And more than enough to flummox an inept ecologist. When a few more of my attempts to land a net on one result in failure, my instructor takes pity on me. She calls me to see a white-claw she's nabbed. Up in the air, claws squeezed together to keep them from mischief, it looks distinctly annoyed. And pretty similar to the American signals. But it's a lot smaller, and a lighter brownish colour. The white-clawed's name comes from the whitish patches around the joints of its claws. They aren't particularly impressive or attractive to look at, but that's not the point. Not every endangered species is a looker.

Catching these little guys in upland rivers requires skill. But luckily for me I wouldn't need it. The rivers in Oxfordshire are deeper and slower. That meant I could set traps.*

---

* The risk with traps is of them being dislodged from the river bed and washed downstream. Fastened to the bank by a rope and peg, a trap basically acts like a big cylindrical kite if caught in the flow. It is possible to set traps in fast flows but you need to select the location with care. A back eddy, or pool right by the bank, can work. But even then you often return to find the thing bobbing and straining on the surface, metres downstream from where you set it.

The idea was that I would set pairs of them in predetermined positions across a carefully selected sample of rivers. Of these, one trap would be a big, plastic, commercially available signal crayfish trap. The other would be home-made out of fine wire mesh and cable ties, specifically sized to catch and retain the smaller white-claweds. Making a load of those took a while. But that was OK, because I also had to get and renew a bunch of licences. I needed a licence to trap and handle the endangered white-claweds, and a trap-licence tag to put on each and every trap, as well as yet another licence to enable me to release, once captured, the invasive signal crayfish.* Once all was ready, I arranged access permissions with landowners on rivers all across the Upper Thames catchment and headed out to hassle crustaceans.

It was only then that I discovered that my carefully

---

* For all non-native species (including American mink, grey squirrels, signal crayfish and many, many others), it is illegal to release one into the British countryside, regardless of how it came to be in your possession. If you've just caught a creature from a massive wild population it can feel daft to be told that you can't now put it back where you found it. But then the authorities do need to be able to prosecute folks caught releasing potentially damaging species into the wild.

devised plan had one serious drawback: it was ridiculously optimistic.

In six months I captured a grand total of no white-clawed crayfish whatsoever. The stretches of river where they had been were either now stuffed full of signals, or eerily empty. In retrospect this shouldn't have surprised me. Signals do not play nicely with white-claweds. They are larger and more aggressive, they breed more quickly and reach far higher densities. They easily out-compete the smaller crayfish for food and refuges. They also directly catch, dismantle and eat them. But the empty stretches were puzzling. These were tens of kilometres upstream of the nearest population of signal crayfish, and should have been safe. They should still have had white-claweds. But instead they contained no trace of either species.

There was a good reason: crayfish plague. This fungal disease is carried by the signals. They are largely immune to it, but it is lethal to white-claweds. I spent all that field season disinfecting and drying my equipment after each and every use to avoid inadvertently exterminating a protected species. But I needn't have bothered. The white-claweds were already gone. Maybe somebody's dog took a dip downstream and jumped in further up. Or perhaps some fishing or canoeing gear picked up weed containing the plague and transferred it. In any case the job was done. At the end of six months in the field I had little to show except the conclusion that white-clawed crayfish probably could never coexist with signals, and there was no point trying to find places where they did.

At least this simplified the remainder of my study. We now needed to discover if there was any way to deal with signal crayfish, or any benefit to trying. This became the aim of my remaining two years. But it too was ambitious. Globally nobody had ever managed to eradicate any species of invasive crayfish from anywhere. Researchers had attempted a large and inventive series of potentially genocidal interventions: manual removal, which means using traps and manpower to collect as many as possible; electrofishing, which uses a giant electrode to attract and stun them before removing them by hand; biological control agents, meaning releasing predatory fish species to chew their way through the population; micro-bial insecticides, such as fungi, and diseases; dewatering entire lengths of river until the crayfish die; destroying their habitat; applying biocidal chemicals to kill them; and any and all of the above in new and exciting combinations. Several projects, for example, tried draining ponds and river sections and spraying chemicals on the exposed crayfish burrows. To be fair, some of these biocide approaches did do the job, but only ever in small ponds and short sections of river. And the downside, of course, was that everything else living in the water also died; and even then, in a few instances, the crayfish re-emerged, blinking, when the toxins had cleared. Turkish crayfish, rusty crayfish, red-swamp crayfish, signal crayfish – you name it, the invasive species in question has uniformly laughed off any practicable attempt to deal with it. For the vast majority of animals on the planet this sort of concerted campaign would wipe them out. But not invasive crayfish.

The closest anyone ever came to a widely applicable solution were two separate projects trialling manual removals of signals and rusty crayfish.* In both cases the researchers stopped before they succeeded but reportedly made a large dent. And both suggested that had they continued and got the crayfish numbers right down, predatory fish might have been able to mop up the rest. So manual removal, I decided, was my weapon

---

* For the avoidance of doubt, this is a species (*Orconectes rusticus*), not an invertebrate that needs oiling.

of choice. But my problem was that I was dealing with rivers, not lakes. Whereas lakes have a finite population, and a low chance of being recolonized, any section of river can be quickly reoccupied from upstream or downstream. So I planned an experimental crayfish removal and migration experiment, with two objectives. The first was to see if six months of trapping could meaningfully reduce the signal crayfish population. The second was to see if any such reduction would cause more signals to flood in, undoing all our good work.

The good thing about working on an abundant species is that you can design a proper study. If you work on polar bears, you expect a sample size of a few individuals. But I expected to catch lots of signal crayfish, and so could try a pretty intricate (at least in field ecology terms) experiment. We found four similar kilometre-long stretches of the river Windrush in Oxfordshire and divided each into three zones: a 250-metre upstream zone, a 500-metre middle zone and a 250-metre downstream zone. At all of the upstream and downstream zones we captured signals, marked them – by piercing a harmless hole in the chitin of their tail with a needle – and put them back where we caught them. From two of the middle zones we also marked and returned. At the other two we tipped the signals we caught into buckets, took them to the lab and humanely destroyed them by freezing.* (We gave bags of frozen crayfish to a local wildlife park to be defrosted and fed to their otters. The otters loved it.)†

The answer to the question of whether we could reduce the signal population was no. After sixty days of hoicking signal

---

* Freezing, down to minus 20 degrees, was the approved method for humanely killing crustaceans. Effectively this mimics what happens to crayfish in winter in the wild. When the water gets cold they go torpid, and their nervous system shuts down. For this reason it is practically impossible to catch crayfish when the water is less than 8 degrees or so. Keep on dropping the temperature below zero and eventually they die, but they will have long since stopped responding to anything. Freezing is thought to be completely painless.

† Wild otters are the absolute masters of crayfishing. On the river Glyme in Oxfordshire is a pool above a weir and next to this pool you can routinely find big mounds of crayfish carapaces and claws, several feet across and inches deep in bluish remains, because the local otter, whenever peckish, simply dives down, grabs a crayfish and eats it. This, by the way, is why under no circumstances should anyone go crayfishing near otter holts or indeed where there are populations of water voles. The water voles and otter cubs will go into traps after the crayfish. The water voles will get stuck and drown in almost any trap. If the traps are made of rope (which they shouldn't be), they risk entangling the cubs.

crayfish from our two removal zones we had extracted a whopping 5,570 of them. But this succeeded only in dropping the average number we caught from six per trap down to about five. There were plenty left.

The answer to our second question was that removing signals – thankfully – did not increase the rate at which others dispersed in to replace them. The percentage of the upstream and downstream crayfish that made it to the middle zones was exactly the same where we were marking-and-returning as where we were removing crayfish (and where the densities were, as a result, a bit lower). In both cases, around 4 per cent of the upstream and downstream populations moved in. We did find, though, that the immigrating signals went about 50 metres further if moving into a removal rather than a mark-and-return zone (an average of 239 metres, not 187 metres). To understand why, picture a river bed as an obstacle course comprising big, territorial signal crayfish. Other big crayfish can eventually bully their way through. But small crayfish, or those that have lost a claw,* risk getting beaten up or eaten. (Signal

---

* Crayfish lose claws the whole time. They can bloodlessly eject a whole limb if it suits them (e.g. if it gets stuck in something, or

crayfish are not fussy about their food. And cannibalism is fair play. If you don't want to be eaten, you shouldn't be small and within reach, OK?) So what we had done in the removal zones was reduce the numbers of the very biggest crayfish (the most easily caught), and thereby thin out the obstacles. When immigrating signals reached our lower-density areas they suddenly found less impediment to their progress.

We learnt something useful but failed to make much of an impression. So in the third year of the project we really went for it. We had been using one trap per 15-metre length of river. Now we would have two traps, on opposite sides of the river, stationed every 5 metres. No crayfish would ever be more than 2.5 metres from a trap. And we reduced the lengths of river we were working on to 100 metres each. So now we had four 100-metre sections of river, two designated as removal zones and two as mark-and-return zones, each intensively trapped for eight consecutive days every month for six months. That, we felt, would show 'em.

---

grabbed by a predator), and then regrow it. It was common to find bits of crayfish claw in our traps, or monstrous crayfish sporting silly-looking, tiny appendages. While losing a claw doesn't kill them, it puts a crayfish at a bit of a disadvantage in a fight.

Having closer trap spacing meant we could study the signals' fine-scale movements. The river we chose was so slow that we could position traps to about a metre's accuracy. So if a crayfish was caught in 15A (70 metres down, nearside bank)* and had previously been in 20B (95 metres down, far bank), we could work out the distance it had gone to within about 2 metres. In a field setting with actual wildlife that's pretty good. The data would tell us a lot about the effect our trapping was having, but meant that the animals would have to be individually identifiable – which signal crayfish aren't. So we hit on the idea of writing the trap number they were caught in on the side of each of them. Then we could put them back and see where they turned up the next day. This meant, too, that in the removal sections we would have to return at least some of the crayfish to the water. So we planned to mark and return crayfish from every other trap, and take

---

* The trap numbering always messed with my head. 1A and 1B were at 0 metres, so 2A and 2B were 5 metres further down. At 100 metres it's trap 21A and 21B. So we had 42 traps per 100 metres. And we trapped two 100-metre lengths on any given day, so we were checking 84 traps a day. Somehow those numbers never look right to me.

the rest away. Once a crayfish had been recaptured, it could go into the removals bucket.

Running this project took four of us, working in teams down each side of the river: Rosie (who to this day remains adamant that crayfish are attractive), as my field assistant, and two volunteers. Rosie and her volunteer worked the line of traps on one side of the river while my volunteer and I took the other. We spent the next months catching, writing on and removing bucket after bucket of crayfish. The work itself was a bit aggravating. The best crayfish bait is fish – fish heads if you can get them, or otherwise a plastic bait box with a chunk of cheap frozen fish in it. (The box stops the fish from being guzzled in the first hour and the crayfish then spending all night trying to find a way out of your trap – which they often do.) I hate the smell of fish. I always have. And now my field clothes, all my gear, my clipboard, my car, my hands, and seemingly everything I possessed or touched permanently smelled like a seal's breakfast.

Writing on crayfish, admittedly, is fun – for about five minutes. After that it's a pain. You have to dry them with a towel so the permanent marker sticks. This is almost impossible if it's raining. Their carapace is abrasive and spiky, so the nib of the pen never lasts, and your hands are permanently cut and sore. And sometimes a crayfish gets lucky and grabs you in that soft bit of skin between your fingers. If you try to pull them off, they tighten their grip. Some painful experimentation revealed to us that the best way to get a crayfish to let go of you is to dangle it from the claw with which it's holding on. This is excruciating, but the crayfish soon gives up and drops off. So trying to write legibly on an attacking crayfish is something of an art form,* and one which is further hampered by their constant tail flicking. Tail flicks, as with the white-clawed crayfish, are how they escape predators, and in the water does lead to some beautiful behaviour. On land the best you can hope for is that it sprays river water over your face. The worst is . . . unpleasant. If you're holding a pregnant female, you get spattered with eggs. If you're really unlucky the eggs have hatched and the female is carrying hundreds of millimetre-long young tucked under her belly. One false move and you are combing

---

* An art form at which my assistant, Rosie, was a genius. She could handle four crayfish at once, each thrust between the fingers of one hand. She looked like an avant-garde Freddie Krueger.

them from your facial hair. And sometimes, early in the season, male crayfish carry long tubes of sticky sperm on their underside. Let's not dwell.

Anyway, the results of our third year were genuinely spectacular. Albeit for the wrong reasons. This time we removed a total of 6,181 crayfish from two 100-metre stretches. Yes, that is indeed 3,000 crayfish per 100 metres. And this vast haul only lowered the average number we caught per trap from eight down to about six. Our total from all four stretches was 27,354 captures of 15,793 individual crayfish. We might have made a very small dent in the population but we were laughably far from eradicating it.

The worst news is that those 15,793 were nowhere near representing the full population down there. The crayfish we caught were only a proportion of the very biggest ones that we could catch. Smaller crayfish stay well clear of the large ones, quite reasonably not wanting to be rent limb from limb. They can't be caught once a trap is full of big monsters. So the best I can say is there was an immense, unknown population of signal crayfish – ranging in size from a couple of millimetres to the biggest thugs – that was undoubtedly many, many times larger than the 15,793 we actually recorded. And all in a combined length of river shorter than the street I live on. Argh, frankly.

Signal crayfish densities can reach 10 or 20 per square metre of river bed. These densities are probably now typical across the vast majority of Britain's freshwaters. There are something like 250,000 kilometres of rivers and streams in the UK, perhaps representing something in the order of a billion square metres of river and stream bed. These figures are very rough, but you can add to them the untold square metres of pond and lake beds that have a suitable depth profile for signal crayfish. Combined, we are looking at a national population of invasive signal crayfish that could number in the billions upon billions.

If that happened on land, it would be a national emergency.

Imagine thousands of square kilometres of Britain's parks, roads, gardens and play areas smothered with lumbering crustaceans. Imagine stepping from your house and finding hundreds of them on your front lawn. Imagine farms, moors or forests covered with the things. People would be out with spades and nets, dealing with the infestation. But instead this is occurring beneath the surface of Britain's lakes and rivers, and of lakes and rivers across the globe.* It is under the water and out of sight. And so barely anybody knows or cares about it.

And here's the killer question: what are those crayfish eating?

Because signal crayfish eat anything, absolutely anything, they can get their claws on. They shred vegetation, they chew up detritus and rotten leaves, and if something is slow-moving and made of meat, or even fast-moving and inattentive, it is going down a crayfish. In that category are all of our native macroinvertebrates, fish eggs, fish fry, frog and toad spawn. The signal crayfish, present in monstrous densities that the white-claweds could never reach, are munching their way through whole populations of our aquatic species. Nobody seems to have comparative data showing the state of our freshwater communities before and after signals, but we at least managed to discover the effect of trapping them through our own project.

Another of my friends, Alison, worked with us to place

---

* Not always signal crayfish, but take any crustacean from its native waters and let it become invasive elsewhere and it seems disproportionately likely to cause absolute, and often irreversible, ecological mayhem.

colonization samplers in our rivers. The samplers look like futuristic tower-blocks, rendered in miniature in plastic: stacks of shallow circular dishes that sit upright in the river, waiting to receive a rain of invertebrates that are deposited into them by the water. (To stop the signals treating these as a convenient and rather thoughtful buffet we also wrapped a proportion of the samplers in mesh, too small for the signals to get a claw into.) In the places where we were not removing crayfish, the colonization samplers collected about nineteen individual macroinvertebrates of six different types. By comparison, right in the centre of the removal sections, where we got crayfish captures down to one or two per trap, we were collecting up to a hundred individual macroinvertebrates, of about fourteen different types. This evidence suggested that the macroinvertebrate population had rebounded in the comparative absence of signals.

And there was further evidence. The signals' daily movements were roughly 30 per cent smaller where we had lowered their densities. The actual difference was not huge, about 11 metres rather than 16 metres, but the likely ecological explanation is that suddenly they had access to a lot more food. Why roam further when you have all the macroinvertebrates you can eat? We found too that at the removal sites, as the project wore on, the remaining crayfish got progressively heavier for a given body length. Crayfish grow by moulting. They shed their old carapace then quickly form a new one.* The length of their carapace does not change until they moult, but they can put on weight between moults (and use this weight to get bigger, faster during a moult). How heavy they are for a given carapace length is therefore a fair indication of how much food they are getting. Perhaps the removal of the big crayfish released the smaller ones from the threat of bullying, so they could feed more efficiently. Or possibly we were seeing the effect of decreasing crayfish numbers resulting in an increase in the abundance of food (i.e. invertebrates). Both explanations are supported by our finding that crayfish at the removal sites became progressively heavier for a given carapace length, suggesting they were getting good feeds in. We cannot be certain which was happening, but the picture is of a macroinvertebrate community, and a community of smaller

---

* Incidentally, there is very little in life more disgusting than handling a newly moulted crayfish. Until their shell re-forms they are like a cold, slippery and semi-dissolved wriggling jelly sweet.

signal crayfish, simultaneously released from the threat of being got by the biggest signals – and both communities blossoming as a result.

So overall there was a clear inverse relationship between the densities of signal crayfish and the abundance and diversity of macroinvertebrates. This is both a real worry, and some cause for optimism. It shows we can probably bolster our native aquatic macroinvertebrate communities by removing signal crayfish. But wherever we aren't doing this – which is in 99.99 per cent of our rivers – the signals are stripping the place bare.

Where does all this leave attempts to control signal crayfish? Well, it means they are futile, if our intention is to eradicate them. They are here to stay. Get used to it. But at the same time widespread intensive removals would probably hugely benefit an array of other aquatic species. It is also possible, given that the signals seem not to move around quite so much when their densities are lowered, that we could slow the rate of advance of signal populations with intensive trapping. So there are good arguments for control attempts aimed at managing the numbers of signal crayfish.

And there is also another, non-biological, argument.

While working on the river we had a lot of conversations with fishers. Many of them told us that the river used to run clear, but these days it was murky. They blamed the signal crayfish. Our research suggests that they were right. We worked with a team of geomorphologists from Queen Mary's University in London to study the turbidity (a measure of the amount of sediment and general small bits of stuff – or 'spaff', as Rosie terms it – suspended in the water) of our rivers. It was worth a shot, and it paid off. Throughout the day the water's turbidity remained level, but as soon as the sun went down turbidity levels began to rise, especially in the sensors nearest the river bed (where the crayfish were). Turbidity quickly peaked and stayed high until just after sunrise, after which it began to decline back to baseline. There is no obvious geomorphological explanation for why rivers should get more turbid at night. There is, however, an obvious biological one: signal crayfish are nocturnal. There are gargantuan numbers of them chewing up leaves and detritus – causing fine particles to be suspended in the water – and scuttling around, ten legs each stirring up the mud. And one of their characteristics as a species is they excavate immense burrows, with tens of entrances in every square metre of bank. The burrows cause banks to erode and often collapse, vastly increasing the amount of sediment in the water. And lab experiments on signals kept in artificial rivers show clear evidence that their activities can create turbidity. In these studies the number of sediment pulses in the depths of the night, caused by signal crayfish, was almost three times higher than during the daylight.

The evidence stacks up. Signal crayfish have made our rivers emptier of everything except signal crayfish and sediment. Scaled up, the implications of their nocturnal industry could be vast. They could increase flood risk in our towns. Their activities change the ways that sediment is transported and accumulated. They destroy stands of aquatic plants (which would otherwise stabilize banks and mop sediment from the water) and stir up particles that then flow unchecked downriver. They undermine flood-defence works with their burrows. In places where agricultural run-off or mining activities have contaminated sediments, these could be resuspended, with dire implications for water quality. And widespread increases in turbidity block river plants' access to light. This, coupled with the signals shredding them, threatens any aquatic species, be it fish, amphibian or invertebrate, that relies on healthy stands of aquatic vegetation for food and shelter.

If all of this sounds like an ecological disaster, that's because it is. There is literally nothing about the signal crayfish that is good for any aspect of British aquatic ecology or geomorphology. And the very worst aspect of this catastrophic infestation is that it is, of course, self-inflicted. They didn't get here by chance. They were deliberately let go. The signal crayfish established and spread in the UK as an entirely predictable – and indeed entirely predicted – consequence of commercially driven decisions, and a lack of oversight and planning that borders on environmental vandalism.

In 1976 a consortium of enterprises, with enthusiastic support from the British government, had a simply brilliant* idea. Wouldn't it be amazing, they thought, if we could get involved with aquaculture and harvest home-grown crayfish for sale in our (tiny) domestic market, and the (far larger) Scandinavian market. Our native species was too small, of course, and another candidate would be needed. The candidate they selected was the American signal crayfish, known

---

* Note: sarcasm.

to reach a great size and to breed prodigiously. To get the project under way they gave us the hard sell. At one point it was even put about that our native species was extinct (it wasn't, not even slightly). So the argument could be made that by bringing in the signals we would be returning crayfish to British waters. Which would be great, right? Doing the place a favour, really.

And so, in contravention of all of the lessons from the wealth of then-existing case studies of similarly stupid ideas (mink, anyone?), and in defiance of contemporaneous warnings from those boring environmentalists about the likelihood of escapes and the probability of crayfish plague being unleashed, blah, blah, blah, a large number of signal crayfish was introduced from Sweden and released into ponds and lagoons. Success! A British Crayfish Marketing Association was, optimistically, set up to control prices and sell stock. And not for these guys the small and well-quarantined initial test site. No precautionary principle here, nor any well-studied cost–benefit calculation, or cautious roll-out only once it had been proved beyond doubt that no ill-effects could occur. Nope, if you're going to do something, move fast and break things. Signal crayfish were quickly introduced to a lot of places: 110 known farms between 1976 and 1990. And they broke our rivers.*

They got out. Of course they did. As I said, they always do. Signals can survive for a long time out of water. They are proficient at climbing through or over pond outfalls and into rivers. And once out, they establish quickly. Signal crayfish farming rapidly became unprofitable due to a lack of domestic demand, and further releases were banned in 1996 under emergency legislation to protect the white-claweds. But by then some of the more unscrupulous commercial crayfish catchers had realized that they could charge riparian landowners for 'nuisance crayfish removal' services, and then sell

---

* By the way, you may recall that I said that otters are absolute masters of catching and eating crayfish. This is true. But I did a back-of-the-envelope calculation to see how much otters were likely to help with keeping the numbers down. I reckon an otter would have to eat about 150 crayfish a day every single day of its life to have much of an impact on crayfish population. I don't know how many crayfish an otter can eat before throwing up,* but I suspect that it's in the low tens.

　* I did a quick online search to see if otters *can* throw up. I am nauseated to report that the answer, as evidenced by an educational YouTube video, is yes.

the supply of live signals thus obtained to restaurants. And with profit on the horizon, signals mysteriously started appearing all over the place – in point-invasions suspiciously close to road bridges over rivers, where a bucket of crustaceans could be furtively dumped into the water in the early hours.

The rest is history. White-clawed crayfish, except for a few small enclaves, are done for as a wild species in Britain. The plan to create a fishery failed in one way but was brilliantly successful in another. It did indeed create a gargantuan supply, as intended, just waiting to be exploited. There are livings to be made. And beyond the host of good ecological reasons for rolling out trapping across as many rivers as possible,*

---

* There are understandable, but not insurmountable, reasons not to. The first is that the licensing would be time-consuming for the Environment Agency and would cost money that could be used on competing priorities, of which they have many. Licensing is needed because a crayfish trap of the wrong type (those made from rope, not plastic) can catch and drown otter cubs. And crayfish traps of any type in places with populations of water voles could catch and drown the voles. Traps with holes in the top might allow the voles out, but also let all the crayfish out too, rendering them useless. So basically somebody needs to be in charge of licensing. That duty would be an

there are enough American signal crayfish in British rivers to keep every man, woman and child in protein-rich, delicious crayfish meat in perpetuity.

But we're not eating them.* The hidden horde goes almost entirely unmolested. Unbelievably, the majority of the crayfish in British supermarket sandwiches is imported from China. Some crayfish meat, I'm told, is even imported, rinsed once, then resold as 'British produce'. Chinese crayfish meat is cheaper, because dismantling a crayfish requires human labour, and such labour in Europe is expensive.

Thanks to economic realities the four-decade-long ecological nightmare unleashed by one rush of enthusiastic money-making, supported by misguided governmental policy, has yielded none of its intended benefits. Because we are determinedly not catching and selling crayfish.

Instead of the consortium getting rich, society is left with

---

unwelcome burden for any grotesquely underfunded environmental agency.

* Well, except for Crayfish Bob, known also as Lennie Usculus, who I believe runs 'crayfish boils' in London. When I met him he was already a star of Oxfordshire crayfish-catching. I hope he still spends summers on his narrowboat, hauling in the signals.

the bill for its failings. A very conservative estimate* of the money expended in managing signal crayfishes' populations and repairing their impacts on flood defences is £2 million per annum. And the irony is that the damage, material and ecological, could be reduced and managed if we were willing to perform the one activity that was the whole point of the releases in the first place: catching crayfish.

---

* Given this figure is from 2010, before our work on crayfish and sediment was published, it's actually most likely to be a colossal underestimate.

7

Elegy for a River

There is a sacredness in tears. They are not the mark of weakness, but of power. They speak more eloquently than ten thousand tongues. They are the messengers of overwhelming grief, of deep contrition, and of unspeakable love.

WASHINGTON IRVING

STREAMS SPRING AND TRICKLE. THEY MELD AND COMBINE. THEY form rivers, narrow and rushing, then broad and powerful, then sluggish and brackish as they wash through estuaries and deltas and out into the sea. River waters flow in channels beneath snow-covered banks, or beside cool, lush grasses and trees, or through fragile ribbons of the only green to be found in brown, dusty lands. But in every country and continent, rivers bring abundance. Our very first civilizations grew from river valleys, nourished by their waters. And ever since, they have fed our hunger, met our thirst, conducted our trade, cradled our lives.

And yet we have allowed them to be diminished. We have inflicted losses and presided over changes that have harmed wildlife and humans alike. If I thought anyone would make it past the first page, I could fill volumes with the mammals and birds exterminated, the global amphibian carnage from pollution or chytrid fungus, the plummeting numbers of insects and plants; or the effluents released – from industry, mining, oil prospecting – that contain heavy metals, toxins, acids;* or the flows diverted, over-extracted or dammed, depriving now-thirsty lands downstream; or the invasive plants and animals – horticultural escapes, too-big-to-feed pets, or well-meant but stupid releases intended to make an attractive addition, a quick buck, or act as biological control – which have established and spread to throttle flows, obliterate light or chomp native populations to death. But in the end it's all the same thing. We continue to strip our rivers of their flourishing finery. We reduce them to mere flows of water, even those adulterated and unsafe.

Conservation can be pretty depressing. But conservationists by their nature are optimists. Our response has always been to take a breath, renew our enthusiasm and treat every

---

* I was told of one particular leak that occurred in the UK where a factory accidentally released a shot of hydrofluoric acid into a small stream. For months afterwards the water ran crystal clear, because it had been sterilized of life. Above the water the plants on the banks were scorched and dead, as though they had been set ablaze. This sort of thing happens the whole time, everywhere.

loss as a problem to be solved. The task has grown exponentially larger, but we have never believed it to be insurmountable. Because beneath everything, justifying all the
dedication and effort, was the deal. We accepted the risk of
failure so that we could be in with a shot at the solution.
And the reward for finding it was always that golden chance
to return to the field, decades on, and gaze with satisfaction
at the results of our labours.

Well, it is now a decade on. Let's see how that deal worked
out. For a total of eleven years I studied two conservation
problems in which were enmeshed four principal species –
two native, two invasive. The work, which ended in 2013,
resulted in one thesis, seventeen published academic papers
and six book chapters, as well as contributing to several editions of *The Water Vole Conservation Handbook*.* The results
and recommendations are ready and available for anyone who
might be interested. My small contribution has joined with
those of many other researchers on just these two topics and,

---

* Scoff ye not. I'll have you know it's something of a bestseller in the
water vole conservation world. Which I admit isn't saying a huge deal.
But, look, I'll take whatever plaudits I can get.

together, we have created solutions. But in response to those solutions, the respective statuses of Britain's water vole and crayfish populations remain almost completely unchanged.

For all the real-world, on-the-ground, species-saving impact that my research has had, I could pretty much just have spent my time bumbling amiably around the British countryside.* My research achieved almost nothing of practical value. And in this it is far from unique. For each of those eleven years there have been equivalent investments by innumerable researchers working on vast arrays of plants and animals in different ecosystems across the globe. Some have meticulously unpicked the tangled ecology of the green and growing, or of the small, scurrying and multi-legged, of the tiny bities and the entirely innocuous. Others have engaged with the massive, toothy and variously ferocious animals that need swathes of land or vast tracts of ocean just to have space to exist. We have worked with people to find ways to mitigate shootings, poisonings, snarings, and to protect habitats from logging, overfishing, or conversion to food crops or oil palm. We have

---

* OK, I clearly *did* spend my time bumbling amiably around the British countryside. But I'd like to think it was to some purpose.

striven to understand how climate change will shape our frag-
mented ecosystems, what new challenges it will introduce. We
have studied how to connect the bits of what is left, how to
work around and with humans. Over hundreds upon hundreds of thousands of collective years we have collated and
analysed evidence, tested hypotheses, and weighed judge-
ments and conclusions. We have defended our work against
the intellectual scrutiny of our peers. If shown data that con-
tradict our views, we have swallowed our pride and changed
those views.* We did all of this so that when we presented our
recommendations, it would be in the knowledge that they
were a solid basis for the actions needed to preserve species.

And with very few exceptions, those recommendations
have been met with a shrug and a quick return to business as
usual.

In the last fifty years alone, humanity has extinguished

---

* Which isn't to say we did it with a good grace, of course. It is
deeply unpleasant to have to say you were wrong about something –
especially if you happen to have a back catalogue of publications
on the subject. We do it, though, because a willingness to change a
viewpoint in the face of evidence is a fundamental part of what makes
a scientist.

60 per cent of the world's populations of wild vertebrates. Those years quite neatly map on to the period from the inception of modern conservation science to the present. Had we not been working so hard for so long, perhaps far more damage would have occurred. But it is also clear that we have completely failed in our goals. We put our faith in the unspoken bargain that when we knew enough to solve the problem we would, well, get on and solve it. We have only recently realized that such a bargain, being unspoken, was easily disregarded. Even when, twenty years ago, I started working on water voles, relatively few conservation research projects had ever been truly settled. My work came along at just the right time to discover what we can expect when a species is moved from the box labelled 'More research needed' to 'Sorted, so let's get cracking.' The answer is this: nothing.

That we have not saved water voles is revealing. They are famous, beloved and harmless. Any concerted restoration attempt would have a high likelihood of success. The result will not negatively impact on any other native species* or

---

* I suspect, in fact, that we would get a massive thank-you from many of our native predators who (to anthropomorphize for a moment) are

endanger human lives or livelihoods. I'm sure people would support the work, and feel happy knowing it had been done. Yes, there are ethical questions around our duties with respect to the American mink, and whether it is acceptable for us to eradicate one species for the benefit of others. While I in no way wish to diminish the importance of those discussions, let's be clear: they have never been the main impediment to conservation action. The main impediment is the cost. It's the tens upon tens of millions of pounds that we simply do not have to spend on conservation in general, let alone for one species. Research gets done because research is cheap, and it looks good. It looks like a problem is being dealt with. But that's an illusion, because knowledge without action is just knowledge. There is little appetite among the authorities to supply the quantities of cash required to pursue an outcome – and so nothing, in real terms, is achieved.

Society views conservation as not desirable enough to fund. So perhaps we have a marketability problem? It is tempting to think that with more appealing products we

---

probably quite annoyed that a bunch of gluttonous mink have deprived them of an ongoing vole buffet.

should be able to get what we want. Most of my colleagues have given up working on single-species-focused conservation because, well, there isn't a lot of point, is there? These days the only groups of conservation scientists who do still work on the ecology of single species are those whose species are properly cool: the ones that capture the public imagination, that are used by advertisers to sell products and which politicians simply cannot ignore. If a vole goes extinct some people will feel sorry. But tigers, or snow leopards, or polar bears? These animals would be lamented by everyone. Charisma. Now that's marketable.

This is not new thinking. Conservation has a long history with charisma. Charisma, and the desire to protect amazing animals and landscapes, is what got the first US national parks created back in the nineteenth century. And it's still being used. Big animals need big habitats. To secure a population of magnificent jaguars, painted hunting dogs or snow leopards means protecting the thousands of square kilometres they require as home ranges. Those square kilometres also happen to be full of the other animals and plants that live in that space with them. So we target 'umbrella' species, knowingly leveraging their popularity to shelter everything else we want to preserve. But in pulling that lever, to gain the funds, influence and impact we need, we must be strategic. We need to treat the preservation of global biodiversity as though we were planning a war. We have researched where the world's largest biodiversity bang-for-your-buck can be found and which charismatic animals' ranges overlap these areas. We have maps showing which countries and governments might be persuaded to conserve which populations of which species. So we know where to direct our limited reserves. We won't waste resources targeting countries too poor, corrupt or war-stricken to conserve habitats. We don't waste effort, these days, on the ecology of smaller animals, however fascinating. We think big, and work with governments and organizations that can guarantee protection of the habitats we need. We research how to connect these up, consider the wider picture. We focus on protecting just a very few flagship, charismatic species, and in doing so aim to reap far larger conservation rewards. That, in a nutshell, is the charisma plan.

So, which are the most charismatic animals? And is the plan working? In part the answer to the first question depends on who you ask (e.g. their outlook, cultural background, or nationality) and what information you give them (e.g. if the animal is rare or threatened, it's automatically perceived as being cooler). But big cats and primates are always bound to feature highly, with tigers way, way out at the top of any list. A recent study, combining information from sources including animal representation on the covers of Disney and Pixar movies,* came up with a top ten: tigers, lions, elephants, giraffes, leopards, pandas, cheetahs, polar bears, grey wolves, gorillas. These have a fair claim to be the world's most beloved wild animals. And they do indeed get the lion's share† of attention and conservation effort. But unfortunately this is where the plan breaks down. Because if you look at the conservation statuses of those top ten, they're dire. And actually far worse than most. Except for the wolves, they are all categorized by the IUCN (the global body tasked with listing the status of our wild species) somewhere between Vulnerable and Critically Endangered.

Charisma has unexpected consequences. Relatively rich people in Western societies – people whose governments and

---

* Never let it be said that conservation isn't down with the kids.

† Life goal: leave no pun unturned, no matter how appalling.

personal donations tend to fund conservation – mostly don't live alongside charismatic animals. And if we did, we'd almost certainly kill them. (Remember Lilith the lynx, shot under a Welsh trailer? That's a lynx: a medium-sized species of cat that weighs somewhere between 18 and 30kg and which has, to my knowledge, never harmed anyone. If she'd been a tiger they'd have hit her with a bazooka, just to make sure.) Outside of zoos we almost never see the real thing. But our daily lives are still filled with them. We grow up cuddling them. Excluding teddy bears, 49 per cent of all soft toys sold on Amazon in the USA represent one of these top ten. In a single year, 800,000 *Sophie la girafe* baby toys are sold in France, eight times more than the number of actual, living giraffes in the whole of Africa. And every week we all encounter something like thirty individuals of each top-ten animal virtually, in the form of toys, TV programmes, movie posters, imagery, statues. And being surrounded by an abundance of charismatic animals seems to have had a counter-intuitive side-effect. Half the public, when asked, say these animals are not threatened. This is the exact opposite of what we would expect, given how well known they are. It is impossible to prove, of course, but the vast virtual populations that surround us seem to tell us that all is well. We perceive the animals as common. And this masks the true risk of their extinction from our minds.

There is a further consequence of charisma, embodied in a disturbing fact I have already mentioned. The principal cause of death of individuals among all of these ten animal species is direct killing by humans. Sometimes this is self-defence, or retaliation for crop loss or livestock predation. After all, living with a real elephant is less fun than with a cuddly one. But many of the killings are to feed the global trade in wildlife products, or for pleasure, or both. That's a whole tale in itself, but suffice it to say that charisma is not a conservation panacea. Our most beloved wild creatures receive more attention, conservation effort and funding than the rest, yes. But that funding remains an amount wholly inadequate to save them, or the wildlife that shares their ranges.

So how on earth are we meant to protect that wildlife? Conservation has been described as a crisis discipline in crisis. And at the heart of that crisis is money. Action to preserve wild species comes at a price. For an individual species the cost might be in the tens of millions. But what about the global bill? If we could turn up to some meeting of the world's leaders and present our invoice, for how much should we ask to 'do conservation properly'? Well, if a goal of conservation was to reduce to zero the probability of any more species going extinct, consensus is growing that this would require us to set aside and protect half of all the terrestrial territory on planet Earth. And that cost has been estimated as $80 billion every year. To then also protect half of all coastal and marine areas would be another $20 billion. Which means that in total, adequate conservation needs $100 billion of funding per year. Of this we currently receive somewhere between 4 and 10 per cent.

That's right: we need an extra 90 to 96 billion dollars, give or take. Every. Single. Year. In perpetuity. And it's a lot to ask. In most people's terms $100 billion is a staggering amount. If you earned $10,000 an hour, and worked ten hours a day, five days a week, and didn't take a holiday for two thousand years, you would still have earned only about half a year's funding. But from another perspective, actually, $100 billion is peanuts. The USA spends far more than double that on soft drinks every year. It is just three times what the world spends annually on chewing gum. The markets for confectionery and for saving wildlife are obviously pretty different, but the direction of that difference is important. If chewing gum and soft drinks were to disappear tomorrow it would get people angry, but I'm sure they'd survive.* By contrast, if the wild and all of the life-support services it provides were to be destroyed, countless millions – perhaps billions – of human lives would be ruined.

We tend to think of money spent on conservation as charity. It's a good, ethical cause, and makes the world a nicer place. But when the money has to be found by states and governments, it is viewed as competing with human priorities. How can we justify spending billions on nature when funds are needed to feed people, reduce poverty, run schools and libraries, or build railways or roads? In a competition between having water voles and having sheltered housing, the voles must surely lose out. Because we naturally view humans as more important. And I have little argument with that. Our economies, health, industry and social development simply have to come first, and that, apparently, means that saving

---

* And rates of diabetes would probably dive. If soft drinks vanished I reckon average human well-being would increase, and the world would be spared a gargantuan load of plastic waste. But I really don't want to pick that fight right now.

nature, however worthy the aspiration, comes second. It's sad, but this is the grown-up and stark reality of the hard choices with which our governments are presented.

Except it isn't. That framing is misleading. Nature is not a nice-to-have into which finances are sunk for the benefit of a few plants and animals and to please environmentalists. Nature is the reason why human civilizations have always clustered around rivers: it is the fabric that supports every single human enterprise. And every species lost is a tear in that fabric. Yes, there are many individual plants and animals whose disappearance would not have an obviously dramatic impact on our lives. Water voles, unfortunately, are an example. I personally would be heartbroken, but the vast majority of the world's population wouldn't notice. And yes, there are other species that have larger and more obvious impacts on humans. Signal crayfish in the wrong place have made British life more expensive. They have increased the chance of homes flooding, and of tap water containing toxic substances. But again, the effect on each of us is small. To focus on the impacts of individual species, though, is to miss the point. Now, more than ever, we must pull back and examine the wider canvas. Because the collective impact on our lives of the existence of all of our wild species is almost infinitely vast. They form the complex of ecosystems that support our ability to keep on eating, drinking and breathing. Species don't do nothing. Species keep our climate stable, our oceans productive, our lands habitable, our waters pure and our crops pollinated, fertile and pest-free. Setting aside for a moment the incalculable value of magnificence, beauty, awe and wonder, and adopting instead a purely practical, utilitarian, anthropocentric viewpoint, every plant, fungus or creature lost, every habitat logged, every river dammed or polluted, is a blow to the ability of nature to support our way of life.

One hundred billion dollars a year is nothing like a charitable donation. It is an essential investment that makes sound economic sense. If global agriculture, for example, were the only possible source of this money, it should willingly stump up* the full sum. Because protecting the habitats of wild pollinators would ensure the continuing annual provision of $235–577 billions' worth of pollination services to our food crops. Similarly,

---

* The expression is in poor taste, given what some types of agriculture have done to our forests.

insurance agencies should definitely get together and invest $5–10 billion every year to protect coastal habitats, because this money would save them $52 billion annually in claims for coastal flooding – and, incidentally, prevent misery for all the people whose homes and possessions would be wrecked in the floods. The seafood industry should invest because $5–10 billion spent on marine protection should increase their profits by $53 billion a year. The logging industry is worth $300 billion a year, but worth nothing once the trees are gone. They should invest $80 billion every year to keep their entire source of revenue in existence.

We should all invest* because the combined value of all

---

* The perennial question is *how* we as individuals can make an actual difference. There are, or course, a lot of answers. First and foremost there is no way that conservation issues are going to get globally solved without our governments stepping up and shouldering the responsibility. Our most important actions must, therefore, be to continually pressure our representatives to ensure that our desire for a habitable planet is represented in every last one of their decisions. We need to be willing to vote for that, and vote against those whose world view remains steadfastly 'business as usual'. For me, at this point, old party allegiances are beginning to look redundant. And there are, of course, a lot of non-political actions we can take, too. I have suggested a few at the end of the book.

the ecosystem services we lost between 1997 and 2011 is estimated at US$4–20 trillion. That's the cost of the lost clean water, pollination, crop productivity, pest control and everything else nature contributes to human well-being and economies. Not in total, but every year. The annual price for offsetting a loss of $4–20 trillion dollars is a paltry $100 billion. For about a third of what America spends on carbonated, canned and flavoured water, we can save the natural world and safeguard our lives. This is a tiny sum that could be easily found by the world's governments and corporations. And actually it won't cost the global economy a penny. Inaction, however, will. A recent report calculated that business as usual, at a conservative estimate, would result in a net loss to the global economy, by 2050, of US$9.87 trillion. But enacting global conservation measures would result in a net gain to the global economy of US$0.23 trillion.

And if economic arguments don't do the trick, there is clear, and mounting, evidence that the destruction of biodiversity is firmly linked to an increased risk of global pandemics. Zoonotic diseases (infections that pass from animals to humans) are emerging ever more frequently – because of human encroachment into wild areas, and the global trade in wildlife. Creatures, and their body parts, are being globally transported and thrust into new contexts, where people and livestock have little immunity to the pathogens they carry. The names of the emerging diseases are horribly familiar: swine influenza, avian influenza, Ebola, SARS and, of course, COVID-19. Even if we remain unswayed by the prospect of benefits to the global economy, clean air, pure water, food production, or the preservation of the magnificence and beauty of nature, the fact is that conservation still directly affects us.* Anyone who wants to avoid the chaos and death of a global pandemic should wholeheartedly support initiatives to preserve and restore biodiversity, and to fund livelihoods for those who live beside it.†

---

* It all gets a bit *Monty Python's Life of Brian* at this stage, doesn't it? 'Apart from the ongoing attempt to secure clean air, water, abundant food, a stable climate, a functioning economy and a reduced risk of infectious disease, all while preserving the beauty and wonder of the natural world and a future for our children . . . what has global conservation ever done for us?'

† This last is vital. Because otherwise those living with biodiversity will often have little choice but to hunt it or sell it, legally or illegally, to scrape together something resembling some quality of life. These guys are the poor, exploited souls at the base of the global trade.

Let me put all this just one more way. We are currently heading for a world that by 2050 is ravaged and impoverished to the extent where its ability to keep us alive is beginning to look dodgy. And in that 2050 scenario our failure to act is costing us, financially, $478.9 billion a year; a figure, incidentally, which does not include the effects of the economic sledgehammer known (at the time of writing) as the 2020 coronavirus pandemic. The other 2050 is my fulfilled dream of thriving life, teeming oceans, and wilds cherished and preserved. It's beautiful, heart-warming, ensures the continuation of human life and livelihoods, and also financially rewards us to the tune of $11.3 billion a year added to the global economy. I mean, oh, the agony of choice.

Put like that it's pretty simple. But while our leaders persist in viewing the needed outlay as a drain on resources we remain in deep trouble. Spend now, and the Earth will meet our needs in perpetuity. To ensure such an outcome is surely what our elected representatives are for.

And if the world does decide to act, it will find the tools it needs already lying on the bench. They are just waiting to be grasped, and wielded. Conservation research has provided the vast bulk of the knowledge we require to protect our wilds. And it has shown us how to go further. If there is to be any sort of revolution in global conservation, then it's clear that relatively affluent countries will need to lead by example. Nations in the global north have already lost much of their biodiversity, and still steadfastly refuse to restore it. While this remains the case we have little basis from which to criticize other, more biodiverse, countries for failing to protect their own natural resources.* We can and should restore our biodiversity. Again it is work we already know how to do. We can rewild, reintroducing lost species and replacing vanished ecological processes. We can make our countrysides greener, more lush, more robust. And in doing so we're likely to reap substantial benefits.

If the term 'rewilding' gives you pause, be reassured. The aim is not to return Britain to some primordial state of tangled,

---

* For example, Brazil's Jair Bolsonaro defended his opening of the Amazon rainforests to economic exploitation by pointing to the hypocrisy of foreign governments – whose countries have profited from the destruction of their own resources, and have little more natural capital left to lose – lecturing him on the topic. He's a pretty terrible human being, but the argument of 'Why should we protect this stuff? You don't' is a difficult one to rebut.

uninhabitable savagery.* While that could be appropriate in a handful of places, rewilding is a lot more sensible than that. Indeed it's far more sensible than a whole load of our current management practices. To show you what I mean, I want to tell you about the monster I encountered at the Bure Marshes. The sound of it was nothing you'd normally hear in a fen. It drew me to my feet with my dinnertime bowl of pasta still clutched in my hands. Until that moment the only remotely similar noise I'd heard on the marsh was a thrumming roar, close above and behind me. That one had me craning my neck, expecting a biplane . . . but instead gawping at a brownish, shining cloud of bees that hurtled overhead and zipped away like some weird, fuzzy spear thrown at the horizon. The swarm vanished over the distant treeline, taking its roar with it. And I went back to my trap setting, filled with the afterglow of having witnessed it. This noise, though, the one that had just interrupted my dinner, was distinctly different. It sounded like an orchestra of hair-dryers, if

---

* If you're English and have difficulty visualizing such savagery, imagine you have run out of tea and the local shop is closed. Horrifying, isn't it?

hair-dryers were the size of a lorry. And it just had to be mechanical. I shoved my pasta aside, grabbed my fleece and started jogging towards the source of it. If I was right, this should be worth watching.

If I was right, the hovercraft was about to fly.

I knew the path to it pretty well. I'd spent a few nights over the past months over there, chatting with the mechanic who'd been hired to build the thing. He was a fabulous, bearded hovercraft fanatic and, like me, had been drawn to the marsh in pursuit of cool things. I was there for the vole conservation, he for the heavy engineering. Usually it was just him at the workshop. That was where he worked, cooked and slept. But when I arrived the scene was bristling with people shouting instructions or driving machinery, all tense. And above everything was the colossal noise of the hovercraft's fans. The sound was insistent, inescapable, dominating. But even so one glance at the beast itself drove it from my mind.

'Oh my God,' I murmured.

Imagine that somebody has welded together a flat slab of a platform out of industrial steel girders. In your mind give it the footprint of a good-sized terraced house. And furnish it with a gigantic grabbing arm, mounted right on the front.

This arm (yellow, battered) has to be sufficiently, horrifyingly huge to be able to manoeuvre a cutting head the size of your living room to the base of a big tree, clasp the trunk, whip a chainsaw through it, hoist the tree into the air, spin through 180 degrees and drop it, trunk, branches and all, into a receptacle on the back of the hovercraft. The receptacle in question needs to be a gargantuan incinerator that's been bolted on to the platform. In your imagination take a moment to ensure that this incinerator (rust-stained, massive) is also terrifying. After all, it needs three days to get to temperature and once going can blast a tree – freshly ripped and dripping from the fen waters – to fine ash in seconds. Now, here's the tricky bit. Put all that together in your mind and hold it there, that awful, beautiful, ludicrous, house-sized tangle of weld, girders, gears and hydraulics. Try to grasp some concept of how much it must weigh.

Now, imagine it floating about a metre above the surface of the ground.

Even having seen it, the image gives me trouble. To stop the hovercraft drifting off and accidentally demolishing anything, it was attached to a JCB digger. Oh, and also to two more big tractors. The vehicles were arranged in a rough triangle, each linked to the thing by about 30 metres of chain. But they could still barely hold it. As I watched, the hovercraft somehow gained some momentum in one direction and decided it would go that way.* It moved ponderously but with enough purpose to tug a tractor, wheels spinning, towards a fenland ditch. At this point people started shouting more urgently, and the landowner – whose project this was and who had been standing next to me – politely suggested that my presence was irritating and that I might like to go away now.

I returned to my dinner, slightly bewildered by this audacious attempt to solve a fenland management problem. Because, yes, that was the whole point of the hovercraft. It was intended as a nature reserve management tool. I mean, for sure, fenlands do need managing. A fen, left to its own devices, wants nothing more than to stop being a fen as quickly as possible. It's

---

*. I once watched a school play in which, for some reason, a St Bernard dog was brought on stage during a scene. It was held on a lead by two sixth-formers. At some point the dog got bored, stood up, looked around and meandered vaguely off into the wings. The sixth-formers got a cheer and a round of applause as they were dragged, still battling, from the stage. That's pretty much exactly what the hovercraft did.

desperately trying to grow up to be a lovely dry woodland. 'Fen' is just a phase it's going through. Unlike upland marshes and bogs, which are fed by the rain and acidic in nature – and which often have little if any association with rivers – fens start out as an expanse of low-lying, open, alkaline water. They form as lakes, held in a bowl of waterproof geological clay into which water flows from rivers or aquifers (subterranean flows of water created by rain trickling for millennia down through limestone hills). Generations of plants grow and die and form layers of peat that accumulate. Elsewhere the rhizomes of reeds intertwine to form floating mats. With time, the combined accumulations of peat and rhizomes can become thick and strong enough to support the weight of animals. The odd, bouncing surface of the Bure Marshes I observed when I first arrived there is symptomatic of water that has recently set out on its way to becoming land.

A fen is an intricate lattice of barely-land cut through with the stillest of waters. It is a mess of habitats rammed together in endlessly ramifying combinations. Submerged deep aquatic plants crowd next to shallow waters, where emergent vegetation breaks the surface to form mats or to thrust into the air, right beside wet banks stuffed with riparian plants that in a mere step or two become fenland meadows. And further still from the waters wet-loving trees – alders and silver birch – take root to form carr woodland.* The whole landscape is an intense muddle. Glorious life exploits every diverse nook and cranny.

But fenlands are successional. The ecological processes that drove them to this point will carry them further, and erase them from existence. In the woods the trees drain down the water. And over decades more typical woodland tree species seed themselves. And very soon the open lake is a memory, represented by a few damp, peaty, sedgy patches in an otherwise unremarkable woodland.

So fenlands are a managerial nightmare. If we want to maintain them we have to find a way of keeping the trees down. In times gone by this was simple because whole villages earned their living from the fens. They harvested reeds and sedge – good for roofing materials – grazed their livestock and cut wood and peat for building and burning. But

---

* 'Carr' is a term for lands that are on their way to becoming dry forests, having once been reedy marsh, and have arrived at a sort of waterlogged, scrubby wooded state. They are typically dominated by alders. They are also dark, squelchy and difficult to move around in.

now the fen-folk are gone. A few reed-cutters still supply thatch for local houses, but these days fenland management largely costs money. Staff are paid to go out with mowers, which deals with new seedlings. But nobody wants to stand waist-deep in stagnant water and cut down trees and shrubs. The health-and-safety implications are terrible. The insects would drive you insane, which is not great if you're holding a chainsaw.* Back when people were desperate for job opportunities, landowners could get the felling done. But now you literally can't pay anyone enough.

Which returns us to the hovercraft. It was built to decimate trees. But to be useful it needed to be able to buzz its way across kilometres of fen, dyke and river, and to pull to a stop right where it was needed. The giant fans used on other hovercraft would not be precise enough for this. Dragging it behind a

---

* My own minor version of this makes the point. I once had just finished tagging a water vole, and was still holding the tagging gun, complete with mounted needle, when a mosquito landed on my arm. It bit me. I didn't even think. I swatted it with the hand holding the gun. The needle – which had already been injected into a water vole's scruff – flashed past my forearm. Another few millimetres and I'd have been off for a stay in a nice, clean hospital.

tractor wouldn't work, as we've seen. Poles could be a possibility, like some sort of potentially lethal punting exercise. But the whole steering/movement issue hadn't been solved by the time I left the marshes. I never did find out whether the hovercraft was used. I'd love to think it had spent time beetling around the fenlands but I've heard no mention of it since.

As much as I am in awe of the enterprise behind that hovercraft, the ecologist in me can't help but believe that a lot of time, energy, money and exasperation could have been avoided by taking a more restorative approach. Extensive fenlands existed far before people ever got involved. They were maintained by natural processes, which we have lost. Beavers were the original managers of those trees. The streams blocked by their dams caused new areas to become inundated, perpetuating the fen. Joining the beavers was a host of native grazing species, all of which cropped the fenlands in patches, preventing tree seedlings from getting established. Over vast, flat landscapes the action of all these different herbivores caused an ever-shifting mosaic of fenlands and wet woodlands. And for a fraction of the cost of a hovercraft, a reintroduced population of beavers in the right place would almost certainly do an effective job. At least some steps have been taken down the rewilding path. Now, and for the last twenty years or so, fen managers have stocked small numbers of highland cattle or konik ponies (semi-feral Polish horses, tough as old boots) on the fens. These animals, in the right density, do a good job of keeping succession in check and maintaining the land as a haven for biodiversity.*

There. A couple of restored species, and a management problem solved. As with everything, it's easier said than done and the implementation would require care, but it's an intellectually and pragmatically solid approach. And one that's compatible with modern people's lives. Rewilding is not some attempt to roll nature back like a computer's operating system to a restore point sometime after the last ice age.† It's far more

---

* Of course the herds themselves need managing, because the difference between an abundant fen and a trampled-like-a-wet-Sunday-at-Glastonbury pasture is stocking density. Historically another lost ecosystem process (known as 'wolves') kept the herbivore numbers in check. But for now let's settle for managing these herds ourselves. There's a time and place for considering wolf reintroductions in Britain, but East Anglia, right now, is not one of them.

† Remember Windows Vista? Returning to the last ice age might be preferable.

pragmatic. Instead it asks what natural processes (predation, pollination, decomposition) have been lost, and whether we can bring them back while working for and around humans. So yes, we can certainly rewild by reintroducing a suite of high-profile carnivores or herbivores to a remote National Park. But we can also rewild a city by encouraging councils to plant trees, or gardeners to seed bee-friendly flowers, or by getting people together to cut holes in their fences to let hedgehogs in. In the latter case doing so would provide our vegetables with natural slug defence, delight our children and help conserve one of our best-loved animals.*

---

* I did the sums on this for one of my funders. My best estimate was that a stable population of hedgehogs – which are increasingly being driven from the British countryside and into towns and villages – living in suburban environments, might need something like 0.9–2.4 square kilometres of good, invertebrate-filled habitat. Gardens are ideal in almost every respect except that they're often inaccessible: tons of good habitat the hedgehogs simply can't get at. This is one minor contributing factor to their dramatic national decline. But it's no exaggeration to say that if everybody committed to cutting a couple of hedgehog holes in their fences and stopped using slug pellets (which make the hedgehogs' main food poisonous – and, ironically, means that we're killing off a great source of slug-snaffling), we could save the species.

It's easy to identify a plethora of modern environmental issues where we are paying large sums to implement technical and infrastructural solutions to problems that have their roots in the absence of natural processes. Flooding in the UK costs taxpayers £1.4 billion annually, with a further £1 billion spent on flood-risk management. But ecological restoration could mitigate many of the challenges. Recent work to re-wet mires on Exmoor has reduced the size of river storm flows by a third. And planting trees in uplands means that water infiltrates the soil sixty times faster than if we leave it as heavily grazed pastures – so less water runs quickly into rivers, meaning their flows are smaller during heavy rains. Putting woody debris into upland streams can also more than double the time water takes to surge downstream. All of these restorations now protect countless homes and businesses downstream. And they can help water quality too. Wetland buffer areas – those stands of tall herbaceous plants, reeds, sedges and rushes beside rivers and lakes that used to exist before we removed them to eke out a few more square metres of cropping space – are effective at removing nitrogen. Stands of drier vegetation (e.g. meadow grasses further up the bank) are good at removing phosphorus. These buffer zones combined with the restoration of upland mires and woodlands would reduce flooding and help prevent soil-associated nutrients from polluting our rivers. These interventions go some way to counteract many of the ill-effects of industrial agriculture (and signal crayfish).*

With some compromise and lots of exertion, we could put in place physical changes that will restore habitats in our uplands and rivers and provide much-needed flood defence for our beleaguered towns and villages. But some of that exertion itself is optional. Much of the work could be achieved through reintroducing recently lost native species, sitting back and letting them get on with it. Yep, it's European beavers again, because they love to engineer rivers. They will happily make those changes for free. Their dams create wetlands, slow water flows, reduce erosion and capture sediment.

---

* The scale of the problem, though, is an issue. The concentrations of agricultural nutrients now running into our rivers are *so* high that to meet Britain's water quality commitments one author estimated we'd have to convert about 80 per cent of our agricultural lands into buffer areas. He stated that we need a radically different approach, of the sort that planners and politicians would currently find 'difficult to conceive' – and suggested we consider rewilding large areas as a solution.

Beavers, restored to Britain, could raise water quality and substantially reduce the chance of flooding in our towns and cities. In trial releases they are also credited with overwhelming biodiversity benefits, including for dragonflies, butterflies, aquatic plants, flowers and trees, and for water vole, fish and amphibian populations. Beavers are already back in Scotland, and, as I write, a five-year study in Devon has just ended, with good things to say about the outcomes. In February 2019 a pair was released into Yorkshire, in the hope of controlling flooding and restoring wildlife habitats, and a further release has followed in Exmoor, with another approved in Sussex.

These are green shoots. A handful of beavers do not a national restoration initiative make. Rewilding remains in its infancy. Knowing what I do about reintroductions, I'd be the last person to suggest they are an easy option. But they are an option, and a practicable one. And far from being esoteric or costly, rewilding is cheap compared with the alternatives. But there remain, of course, obstacles. Among these the largest is that highlighted by the Devon beaver project. Its report concluded that 'the ecosystem services and social benefits accrued [of beaver reintroductions] are greater than the financial costs incurred' but that 'those who benefit from beaver reintroduction may not always be the same people as those who bear the costs'.

And there's the rub. The public loves beavers. In a recent survey 90 per cent of respondents said they would support their reintroduction. Released populations in Scotland and Devon draw tourists, which benefit local businesses. One study estimated that a given beaver release could generate between £380,000 and £4.6 million in local tourism revenue, while the total costs should rarely rise above £78,720. But if the people who bear those costs are different from those who profit, and if the latter fail to compensate the former, then the whole enterprise is in jeopardy. Landowners in Scotland have been reported as shooting beavers on sight, blaming them for flooding their land.* They probably wouldn't if they were see-ing a financial benefit.

---

* It's Schrödinger's wildlife – simultaneously cherished and unwanted, profitable and destructive. The only conclusion you can draw when the same creature can be watched by captivated millions on TV and yet snared as a pest by a gamekeeper – because it is destroying the non-native game species s/he is employed to release into the countryside – is that we are in urgent need of some consensus and consistency.

For me, humanity's brightest prospects lie in conserving existing habitats and rewilding where we can. In terms of technical ability and ecological knowledge, there is little more we need to know. The biggest hurdle is social. If we are to realize any future in which society benefits from protected and restored nature, the costs of doing so absolutely must be identified and those who bear them amply compensated, funded by the profits made and money saved elsewhere. This is possible. In Britain restoring our wilds would not only clean and tame our rivers, but cause a flourishing of forests and grasslands and wetlands that would each make a colossal contribution to sequestering carbon from the atmosphere. Every hectare of regenerated forest can sequester several tonnes of carbon dioxide every year, as can our wetlands. The NGO Rewilding Britain suggests that landowners could be paid between £500 and £1,500 per hectare to rewild forests, and that peat bogs could be worth £200–800. This money would be remuneration for the benefits to the whole population only of the carbon sequestration. The figures do not capture any of the concomitant improvements to biodiversity, pollinator services, flood defence, forestry, fisheries, pest control or tourism. But they could. Conserved, restored and rewilded, our lands and waters can be a haven for wildlife, an economic boon and a bulwark against climate change. We have dangling, just beyond reach, that rarest of prospects: a cost-effective win–win–win.

What is true of British rivers is true globally. If humanity benefits to the tune of trillions from the existence of nature, then the custodians of that nature should be accordingly remunerated. Across the world other rivers run down from the wilds, through villages and towns into cities. They bring clean, cold water. They bring beauty and wonder. They bring prosperity. And if the price for that prosperity is that we must share a little of our wealth with upstream folk, to help set aside a little more room for our wildlands, then I say that's a bargain worth having. But this is not my choice to make.

The choice I make is to keep on researching. But it has been a long while, now, since I worked by water, or even outdoors. My dream was always to study creatures in the wild, to solve and remedy their conservation problems. But that's no longer a dream I can realize. Instead I have moved on, grudgingly following the scent of funding and the hunt for ways to make an impact. These days I sit at a desk. I construct surveys that will allow us to understand human behaviour, how it drives the global consumption of wildlife and whether it can be changed. It's a whole other story, and takes me far from the field, and my first love.

But old habits die hard. Yesterday, on a walk, my feet brought me to water. They carried me down a little-trodden path, along a stream that runs into the Thames. Here dappled sun filters through the trees. It yields enough light for the banks to be bursting and lush, stuffed with greater pond sedge and thick, succulent stands of iris. Out of habit, I crouch at the water's edge, seeking some sign of my voles. But where the earth should be trampled, tiny plants poke insolently up through the mud. And just here is a perfect platform for a latrine, lying unused. And, look, any self-respecting water vole would have hit these irises like a lawnmower, before rolling back to her burrow, replete. Instead their blades wave in the breeze, upright and unconcerned. So I stand and continue on my way. I try to enjoy the sun and birdsong. But there are no water voles here. Not any more.

And there should be, so easily.

I wrote this book as an elegy for a river. It is a lament for the new silences that pool beneath shady margins, the absence of munched reeds and speedy escapes into the depths, the vanished chances of a glimpse into a different world; and for the lurking, chitinous clatter, the shredded plants, bare beds and stirred murk that spread beneath our waters' glistening surface. But I don't want it to be. I think I never really did. Despite everything I find myself holding stubbornly to optimism. So much of what is gone is far from irreversible. And more and more we are hearing public demands for action: on climate, on species loss. And we have solutions, ready-made, provided by decades of research. The only question is whether we can find the will and the courage to implement them.

I think we can.

You see, I have this wild hope . . .

Just one last thing.

If you find yourself standing at the water's edge and wondering how to help, you're not alone. I've mentioned the political necessity of voting against anyone who wishes to kick our collective can yet further down the road — but we can each achieve much through small, individual actions. There are wonderful initiatives and resources to get us involved in our own, mini rewilding projects. We can easily improve the ability of our gardens, yards, balconies or window boxes to support wildlife and wild processes. It's as easy as leaving 'untidy' (aka exuberant) garden areas for wild flowers, amphibians and small mammals, sowing pollinator-friendly plants, putting up bee and insect hotels, and opening small connecting holes in fences. Any and all of these will help, and such patches connect one to the next, forming urban wilds. Good examples of British projects are Naturehood and Hedgehog Street, but there are lots of citizen-science enterprises out there. Then at the county level are Wildlife Trusts, all of which do amazing work, and all of which are criminally short on resources. And nationally, in addition to well-known NGOs such as the RSPB and the Mammal Society, a number of enterprises are engaging with larger-scale rewilding projects. Some of these directly crowdfund money to buy up land to manage for nature. Others like Rewilding Britain inspire and enable change. And then there are global conservation charities, and initiatives that seek to provide sustainable livelihoods for those who live closest to the most biodiverse global habitats. Any of these initiatives, at any level, would be worthy recipients of investments of your time or money. It's always tempting to feel that we have little individual impact. But the cumulative effect of people taking small actions, enjoying the benefits and encouraging more folks to get involved can be vast.

And finally, if you wish to help water voles specifically, you can volunteer as a water vole surveyor with your local Wildlife Trust, or donate to your trust's work. Or you can join the National Water Vole Monitoring Programme, run by the People's Trust for Endangered Species (PTES). And then perhaps you too will find yourself, clipboard in hand, smiling at a pile of chopped plants. Or at the 'plop' of a fast exit. Or at a rustle in the reeds, and the sight of one, tall stem being hauled downward by industrious, unseen paws.

ACKNOWLEDGEMENTS

OK, there are a huge number of folks I want to thank, but I'll keep this short. Lots of people suffered on field projects with me. Guys, I'm sorry. The worst of my ineptitude was borne by Ian Ellis, Rebecca Dean, Merryl Gelling and Rosie Salazar, all of whom compensated amazingly. They escaped with only a few lingering verbal tics and twitches.

This worthy volume represents the melding of two careers. This may or may not be a good thing. But I wouldn't have been any sort of researcher without the continual support of the indefatigable Professor David Macdonald, and all the wonderful folks who work with me at WildCRU. I still have no idea when David actually sleeps. And I wouldn't be anything like a writer without the guidance, patience, encouragement and occasionally, let's face it, gritted perseverance, of the brilliant Catherine Clarke. (Also, this book was her suggestion. Blame her.) Thank you both, I owe you more than I can say.

I'm so grateful to anyone who reads a draft manuscript and responds with as much enthusiasm and commitment as Susanna Wadeson. It made writing the full version a delight. I have no idea whether it's the done thing to thank your editor in the book she's edited, but I'm doing it. So there. Thank you!

And lastly, my family. I can't write without you. I love you.

BIBLIOGRAPHIC ENDNOTES

There's a balance between the level of referencing most academics would approve of (i.e. substantiating every single statement, without exception, until any remaining zest for life curls up and dies) and a book that someone might actually enjoy. I've tried to achieve this balance by ensuring that the key facts are referenced. The only exceptions are well-known water vole natural historical details, which I've left unreferenced – because, well, the book is full of them, and we'd all be here all day. The information can, however, be found in *The Water Vole Conservation Handbook* in its various editions (to some of which I have contributed), and in more digestible form in *Water Voles* (British Natural History Series) by Rob Strachan.

I have ended up citing my own work a lot. This is inevitable but feels more than a bit self-indulgent. To give things their proper context, I made a minuscule contribution to a vast body of conservation knowledge on a dazzling array of subjects. Our individual research projects peek under the carpet in different corners of the house. But, to over-extend a metaphor, the revealed state of the floor is the same in every room. I wouldn't put too much weight on it. Maybe call a builder.

Chapter 1

(Nothing to see here. Seriously, no refs.)

Chapter 2

General note – in this chapter I detail a lot of the evidence of the causes of the water vole's decline, but, as I say, I also left a lot out. Anyone wanting a more intricate account can find one here: Moorhouse, T. P., Macdonald, D. W., Strachan, R. & Lambin, X., 'What does conservation research do, when should it stop, and what do we do then? Questions answered with water voles', in Macdonald, D. W. & Feber, R. E. (Eds), *Wildlife Conservation on Farmland, Volume 1: Managing for nature on lowland farms*, 269–90 (Oxford: Oxford University Press, 2015).

26 Many die immediately: Jarić, I. & Cvijanović, G., 'The Tens Rule in invasion biology: measure of a true impact or our lack of knowledge and understanding?', *Environmental Management* 50, 979–81 (2012).

26 Do rabbits belong in Britain?: Lever, C., *The Naturalized Animals of Britain and Ireland* (London: New Holland, 2009).

27 But ecologically I'd have been wrong: Sandom, C. J. & Macdonald, D. W., 'What next? Rewilding as a radical future for the British countryside' in Macdonald, D. W. & Feber, R. E. (Eds), *Wildlife Conservation on Farmland, Volume 1* (Oxford: Oxford University Press, 2015).

27 a dismay that verged on hysteria: https://www.theguardian.com/uk-news/2017/nov/11/lilith-escaped-lynx-is-killed-over-growing-public-safety-fears and https://www.cambrianwildwood.org/lillith-the-lynx

27 lynx are a natural part of the British fauna: Hetherington, D. A., Lord, T. C. & Jacobi, R. M., 'New evidence for the occurrence of Eurasian lynx (*Lynx lynx*) in medieval Britain', *Journal of Quaternary Science* 21, 3–8 (2006).

29 They are beavers, and so not exactly terrifying: Kitchener, A. & Conroy, J., 'The history of the Eurasian beaver *Castor fiber* in Scotland', *Mammal Review* 27, 95–108 (1997).

31 When the world began to warm again: Hewitt, G. M., 'Post-glacial re-colonization of European biota', *Biological Journal of the Linnean Society*, 68(1–2), 87–112 (1999).

32 The next oldest samples of water vole remains: Brace, S., Ruddy, M., Miller, R., Schreve, D. C., Stewart, J. R. & Barnes, I., 'The colonization history of British water vole (*Arvicola amphibius* (Linnaeus, 1758)): origins and development of the Celtic fringe', *Proceedings of the Royal Society B*, 28320160130 (2016).

32 These three specimens: Ruddy, M., 'The Western Palaearctic evolution of the water vole *Arvicola*', submitted to Royal Holloway College, University of London, for the degree of Doctor of Philosophy (2011).

32 In modern times there remains a genetic distinction: Piertney, S. B., Stewart, W. A., Lambin, X., Telfer, S., Aaars, J. & Dallas, J. F., 'Phylogeographic structure and postglacial evolutionary history of water voles (*Arvicola terrestris*) in the United Kingdom', *Molecular Ecology* 14, 1435–44 (2005).

33 They came from a Greek navigator: https://en.wikipedia.org/wiki/Pytheas#Voyage_to_Britain

35 But over time, the whole thing is stable: Lambin, X., Le Bouille, D., Oliver, M. K., Sutherland, C., Tedesco, E. & Douglas, A., 'High connectivity despite high fragmentation: smart iterated dispersal in a vertebrate metapopulation', in Clobert, J., Baguette, M., Benton, T. G. & Bullock, J. (Eds), *Informed Dispersal and Spatial Evolutionary Ecology*, 405–12 (Oxford: Oxford University Press, 2012).

35 [footnote] The name for what I'm describing: The source text for all things metapopulation is written by the great Ilkka Hanski: *Metapopulation Ecology* (Oxford: Oxford University Press, 1999). It's a seminal work, but best left to the enthusiasts.

39 The omens in question: Jefferies, D. J. & Arnold, H. R., 'Mammal report for 1979', Report for the *Huntingdonshire Fauna and Flora Society* 32, 32–8 (1980), and Jefferies, D. J. & Arnold, H. R., 'Mammal report for 1981', *Report for the Huntingdonshire Fauna and Flora Society* 34, 39–46 (1982).

40 Before 1965 nobody was systematically recording: Jefferies, D. J., Morris, P. A. & Mulleneux, J. E., 'An inquiry into the changing status of the water vole *Arvicola terrestris* in Britain', *Mammal Review* 19, 111–31 (1989).

40 In the 1960s, thanks to the recently formed Mammal Society: http://www.mammal.org.uk/history_achievements (2013).

40 This caused a problem: Jefferies, D. J., Morris, P. A. & Mulleneux, J. E., 'An inquiry into the changing status of the water vole *Arvicola terrestris* in Britain', *Mammal Review* 19, 111–31 (1989).

43 Rob's report came out: Strachan, R. & Jefferies, D. J., 'The water vole *Arvicola terrestris* in Britain 1989–1990: its distribution and changing status', The Vincent Wildlife Trust, London (1993), and Strachan, C., Strachan, R. & Jefferies, D. J., 'Preliminary report on the changes in water vole population of Britain as shown by the National Surveys of 1989–1990 and 1996–1998', The Vincent Wildlife Trust, London (2000).

44 **For example, of 184 early reports:** Jefferies, D. J., Morris, P. A. & Mulleneux, J. E., 'An inquiry into the changing status of the water vole *Arvicola terrestris* in Britain', *Mammal Review* 19, 111–31 (1989).

45 **Two research papers . . . were pivotal:** Woodroffe, G. L. & Lawton, J. H., 'Patterns in the production of latrines by water voles (*Arvicola terrestris*) and their use as indexes of abundance in population surveys', *Journal of Zoology* 220, 439–45 (1990), and Woodroffe, G. L., Lawton, J. H. & Davidson, W. L., 'The impact of feral mink *Mustela vison* on water voles *Arvicola terrestris* in the North Yorkshire Moors National Park', *Biological Conservation* 51, 49–62 (1990).

47 **all but annihilated:** Macdonald, D. W. & Strachan, R., 'The mink and the water vole: analyses for conservation', Wildlife Conservation Research Unit, University of Oxford (1999), and Strachan, C., Jefferies, D. J., Barretot, G. R., Macdonald, D. W. & Strachan, R., 'The rapid impact of resident American mink on water voles: case studies in lowland England', *Symposium of the Zoological Society of London* 71, 339–57 (1998).

48 **These bioaccumulated up the food chain:** Mason, C. F. & Macdonald, S. M., 'Impact of organochlorine pesticide residues and PCBs on otters (*Lutra lutra*): a study from western Britain', *Science of the Total Environment* 138(1–3), 127–45 (1993).

48 **And with suspiciously correlated timing:** Macdonald, D. W. & Strachan, R., 'The mink and the water vole: analyses for conservation', Wildlife Conservation Research Unit, University of Oxford (1999).

Chapter 3

73 **[footnote] 'combat' and 'grazing' settings:** https://www.nationalgeographic.com/news/2014/12/141224-deer-fanged-vampire-animals-science-krampus-christmas/
https://www.wildlifeonline.me.uk/animals/species/chinese-water-deer

Chapter 4

91 **Female small mammals are often territorial:** Wolff, J. O., 'Why are female small mammals territorial?', *Oikos* 68, 364–70 (1993).

97 **males' ranges, too, shrink:** Moorhouse, T. P. & Macdonald, D. W., 'What limits male range sizes at different population densities? Evidence from three populations of water voles', *Journal of Zoology* 274(4), 395–402 (2008).

102 Despite living in a paradise of food: Moorhouse, T. P., Gelling, M. & Macdonald, D. W., 'Effects of forage availability on growth and maturation rates in water voles', *Journal of Animal Ecology*, 1288–95 (2008).

113 It may not matter too much where a given range or territory is situated: Moorhouse, T. P. & Macdonald, D. W., 'Temporal patterns of range use in water voles: do females' territories drift?', *Journal of Mammalogy* 86(4), 655–61 (2005).

114 Where one study showed an intrepid vole: Telfer, S., Piertney, S. B., Dallas, J. F., Stewart, W. A., Marshall, F., Gow, J. L. & Lambin, X., 'Parentage assignment detects frequent and large-scale dispersal in water voles', *Molecular Ecology* 12, 1939–49 (2003).

117 The last thing you'd want to do: Moorhouse, T. P. & Macdonald, D. W., 'Indirect negative impacts of radio-collaring: sex ratio variation in water voles', *Journal of Applied Ecology* 42(1), 91–8 (2005).

Chapter 5

138 Water voles do more of both: Moorhouse, T. P., Gelling, M. & Macdonald, D. W., 'Effects of habitat quality upon reintroduction success in water voles: evidence from a replicated experiment', *Biological Conservation* 142(1), 53–60 (2009).

143 We had previously used LCC: Moorhouse, T. P., Gelling, M., McLaren, G. W., Mian, R. & Macdonald, D. W., 'Physiological consequences of captive conditions in water voles (*Arvicola terrestris*)', *Journal of Zoology* 271(1), 19–26 (2007).

144–5 The more space each vole had: Gelling, M., Montes, I., Moorhouse, T. P. & Macdonald, D. W., 'Captive housing during water vole (*Arvicola terrestris*) reintroduction: does short-term social stress impact on animal welfare?' *PLoS One* 5(3) (2010).

153 Ecologically, they are magnificent: Macdonald, D. W., Harrington, L. A., Yamaguchi, N., Thom, M. D. & Bagniewska, J., 'Biology, ecology and reproduction of the American mink *Neovison vison* on lowland farmland', in Macdonald, D. W. & Feber, R. E. (Eds), *Wildlife Conservation on Farmland, Volume 2: Conflict in the countryside*, 126–47 (Oxford: Oxford University Press, 2015).

153 [Footnote] And the conservation community should absolutely own up: This is a view we first expressed here: Moorhouse, T. P., Macdonald, D. W., Strachan, R. & Lambin, X., 'What does conservation research do, when should it stop, and what do we do then? Questions answered with water voles', in Macdonald, D. W. & Feber, R. E. (Eds), *Wildlife Conservation on Farmland, Volume 1: Managing for nature on lowland farms*, 269–90 (Oxford: Oxford University Press, 2015).

153 [Footnote] Early studies did show a decline in mink signs: Harrington, L. A., Harrington, A. L., Yamaguchi, N., Thom, M., Ferreras, P., Windham, T. R. & Macdonald, D. W., 'The impact of native competitors on an alien invasive: temporal niche shifts to avoid inter-specific aggression?' *Ecology* 90(5), 1207–16 (2009).

158 Swathes of the Highlands are mink-free: https://www. invasivespecies.scot/american-mink-0

158 And there is mink control across much of the south-west: https://basc.org.uk/conservation/green-shoots/green-shoots-in-the-south-west/

Chapter 6

171 There was a good reason: crayfish plague: Holdich, D. M., Rogers, W. D., 'The white-clawed crayfish, *Austropotamobius pallipes*, in Great Britain and Ireland with particular reference to its conservation in Great Britain', *Bulletin Français de la Pêche et de la Pisciculture* 347, 597–616 (1997), and Lozan, J. L., 'On the threat to the European crayfish: a contribution with the study of the activity behaviour of four crayfish species (*Decapoda: Astacidae*)', *Limnologica* 30, 156–61 (2000).

172 Researchers had attempted: Gherardi, F., Aquiloni, L., Dieguez-Uribeondo, J. & Tricarico, E., 'Managing invasive crayfish: is there a hope?', *Aquatic Science* 73, 185–200 (2011), and Stebbing,

P., Longshaw, M. & Scott, A., 'Review of methods for the management of non-indigenous crayfish, with particular reference to Great Britain', *Ethology Ecology & Evolution* 26(2–3), 204–31 (2014).

173 To be fair, some of these biocide approaches: Peay, S., Johnsen, S. I., Bean, C. W., Dunn, A. M., Sandodden, R. & Edsman, L., 'Biocide treatment of invasive signal crayfish: successes, failures and lessons learned', *Diversity* 11, 29 (2019).

173 The closest anyone ever came: Rogers, W. D., Holdich, D. M. & Carter, E., 'Crayfish Eradication', Report for English Nature, Peterborough (1997), and Hein, C. L., Vander Zanden, M. J. & Magnuson, J. J., 'Intensive trapping and increased fish predation cause massive population decline of an invasive crayfish', *Freshwater Biology* 52, 1134–46 (2007).

175 [Footnote] Freezing, down to minus 20 degrees: https://kb. rspca.org.au/knowledge-base/what-is-the-most-humane-way-to-kill-crustaceans-for-human-consumption/

177 When immigrating signals reached our lower-density areas: Moorhouse, T. P. & Macdonald, D. W., 'Immigration rates of signal crayfish (*Pacifastacus leniusculus*) in response to manual control measures', *Freshwater Biology* 56(5), 993–1001 (2011).

181 This time we removed a total of: Moorhouse, T. P. & Macdonald, D. W., 'The effect of manual removal on movement distances in populations of signal crayfish (*Pacifastacus leniusculus*)',

Freshwater Biology 56(11), 2370–77 (2011), and Moorhouse, T. P. & Macdonald, D. W., 'The effect of removal by trapping on body condition in populations of signal crayfish', *Biological Conservation* 144(6), 1826–31 (2011).

182 Signal crayfish densities: Abrahamsson, S. A. A. & Goldman, C. R., 'Distribution, density and production of the crayfish *Pacifastacus leniusculus* Dana in Lake Tahoe, California Nevada', *Oikos* 21, 83–91 (1970); Bubb, D. H., 'Movement and dispersal of the invasive signal crayfish *Pacifastacus leniusculus* in upland rivers', *Freshwater Biology* 49, 357–68 (2004); and Goldman, C. R. & Rundquist, J. C., 'A comparative ecological study of the California crayfish, *Pacifastacus leniusculus* (Dana), from two subalpine lakes (Lake Tahoe and Lake Donner)', *Freshwater Crayfish* 3, 51–80 (1977).

184 This evidence suggested that the macroinvertebrate population had rebounded: Moorhouse, T. P., Poole, A. E., Evans, L. C., Bradley, D. C. & Macdonald, D. W., 'Intensive removal of signal crayfish (*Pacifastacus leniusculus*) from rivers increases numbers and taxon richness of macroinvertebrate species', *Ecology and Evolution* 4(4), 494–504 (2014).

184 The signals' daily movements: Moorhouse, T. P. & Macdonald, D. W., 'The effect of manual removal on movement distances in populations of signal crayfish (*Pacifastacus leniusculus*)', *Freshwater Biology* 56(11), 2370–77 (2011).

185 Both explanations are supported: Moorhouse, T. P. & Macdonald, D. W., 'The effect of removal by trapping on body condition in populations of signal crayfish', *Biological Conservation* 144(6), 1826–31 (2011).

188 In these studies the number of sediment pulses: Harvey, G. L., Moorhouse, T. P., Clifford, N. J., Henshaw, A. J., Johnson, M. F., Macdonald, D. W., Reid, I. & Rice, S. P., 'Evaluating the role of invasive aquatic species as drivers of fine sediment-related river management problems: the case of the signal crayfish (*Pacifastacus leniusculus*)', *Progress in Physical Geography* 35(4), 517–33 (2011), and Harvey, G. L., Henshaw, A. J., Moorhouse, T. P., Clifford, N. J., Holah, H., Grey, J. & Macdonald, D. W., 'Invasive crayfish as drivers of fine sediment dynamics in rivers: field and laboratory evidence', *Earth Surface Processes and Landforms* 39(2), 259–71 (2014).

189 In 1976 a consortium of enterprises: Holdich, D. M., 'A review of astaciculture: freshwater crayfish farming', *Aquatic Living Resource* 6, 307–17 (1993), and Holdich, D. M., James, J., Jackson, C. & Peay, S., 'The North American signal crayfish, with particular reference to its success as an invasive species in Great Britain', *Ethology Ecology & Evolution*, 26(2–3), 232–62 (2014).

190 the probability of crayfish plague being unleashed: Holdich, D. M., Jay, D., Goddard, J. S., 'Crayfish in the British Isles', *Aquaculture* 15, 91–7 (1978).

193 Unbelievably, the majority of the crayfish: https://www.theguardian.com/lifeandstyle/2012/jul/18/why-should-eat-more-crayfish

194 A very conservative estimate: Williams, F., Eschen, R., Harris, A., Djeddour, D., Pratt, C., Shaw, R. S., Varia, S., Lamontagne-Godwin, J., Thomas, S. E. & Murphy, S. T., 'The economic cost of invasive non-native species on Great Britain', CABI Proj No VM10066, 1–99 (2010).

Chapter 7

201 In the last fifty years alone: Grooten, M. & Almond, R. E. A. (Eds), *Living Planet Report – 2018: Aiming Higher*, WWF, Gland, Switzerland.

204 Charisma, and the desire to protect amazing animals and landscapes: https://www.nationalgeographic.com/travel/national-parks/early-history/

204 So we target 'umbrella' species: Macdonald, E. A., Hinks, A., Weiss, D. J., Dickman, A., Burnham, D., Sandom, C. J., Malhi, Y. & Macdonald, D. W., 'Identifying ambassador species for conservation marketing', *Global Ecology and Conservation* 12, 204–14 (2017).

205 We have maps showing which countries and governments: Dickman, A. J., Hinks, A. E., Macdonald, E. A., Burnham, D. &

Macdonald, D. W., 'Priorities for global felid conservation', *Conservation Biology* 29(3), 854–64 (2015).

208 And this masks the true risk: Courchamp, F., Jaric, I., Albert, C., Meinard, Y., Ripple, W. J. & Chapron, G., 'The paradoxical extinction of the most charismatic animals', *PloS Biology* 16(4), e2003997 (2018).

209 Of this we currently receive: Barbier, E. B., Burgess, J. C. & Dean, T. J., 'How to pay for saving biodiversity', *Science 360* (6388), 486–8 (2018).

209 The USA spends: https://www.statista.com/outlook/20020000/109/soft-drinks/united-states

209 world spends annually: In 2015 the world spent $25.8 billion on chewing gum. If you feel compelled to look up this statistic, it's here: https://www.statista.com/topics/1841/chewing-gum/

213 They should invest $80 billion every year: Barbier, E. B., Burgess, J. C. & Dean, T. J., 'How to pay for saving biodiversity', *Science 360* (6388), 486–8 (2018).

213 We should all invest: Costanza, R., De Groot, R., Sutton, P., Van der Ploeg, S., Anderson, S. J., Kubiszewski, I., Farber, S. & Turner, R. K., 'Changes in the global value of ecosystem services', *Global Environmental Change* 26, 152–8 (2014).

214 But enacting global conservation measures: Johnson, J. A., Baldos, U., Hertel, T., Liu, J., Nootenboom, C., Polasky, S. & Roxburgh, T., 'Global Futures: modelling the global economic impacts of environmental change to support policy-making', Technical Report, January 2020, available from: https://www.wwf.org.uk/globalfutures

214 And if economic arguments don't do the trick: Johnson, C. K., Hitchens, P. L., Pandit, P. S., Rushmore, J., Evans, T. S., Young, C. C. W. & Doyle, M. M., 'Global shifts in mammalian population trends reveal key predictors of virus spillover risk', *Proceedings of the Royal Society B: Biological Sciences*, 287, 20192736 (2020), and https://www.weforum.org/agenda/2020/03/biodiversity-loss-is-hurting-our-ability-to-prepare-for-pandemics/

217 The aim is not to return Britain: Moorhouse, T. P. & Sandom, C. J., 'Conservation and the problem with "natural" – does rewilding hold the answer?', *Geography* 100, 45–50 (2015), and 'Rewilding and Climate Breakdown: how restoring nature can help decarbonise the UK', report for Rewilding Britain (2019).

223 And very soon the open lake is a memory: Friday, L. E. & Rowell, T. A., 'Patterns and processes', in Friday, L. E. (Ed.), *Wicken Fen: The making of a wetland nature reserve*, 11–21 (Colchester: Harley Books, 1997); Godwin, H. & Bharucha, F. R., 'Studies in the Ecology of Wicken Fen, II: the fen water table and its control of plant communities', *Journal of Ecology* 20, 157–91 (1932); and Godwin, H., 'Studies in the Ecology of Wicken Fen, IV: crop taking experiments', *Journal of Ecology* 29, 83–106 (1941).

226 These animals, in the right density: *Wicken Fen Vision – The Grazing Programme Explained*, The National Trust, 2011. Available from: https://nt.global.ssl.fastly.net/wicken-fen-nature-reserve/documents/wicken-fen-the-grazing-programme-explained.pdf

226–7 It's far more pragmatic: Moorhouse, T. P. & Sandom, C. J., 'Conservation and the problem with "natural" – does rewilding hold the answer?', *Geography* 100, 45–50 (2015).

228 Flooding in the UK: UK National Ecosystem Assessment 2014, UNEP-WCMC, Cambridge.

228 more than double the time water takes to surge downstream: *Rewilding and Flood Risk Management Briefing* (2016), available from Rewilding Britain: https://www.rewildingbritain.org.uk/assets/uploads/files/publications/Rewilding%20and%20Flood%20Risk%20Management%20briefing.pdf

228–9 Wetland buffer zones: Moss, B., 'Water pollution by agriculture', *Philosophical Transactions of the Royal Society B: Biological Sciences* 363(1491), 659–66 (2008).

229 Yep, it's European beavers again: Rosell, F., Bozsér, O., Collen, P. & Parker, H., 'Ecological impact of beavers *Castor fiber*

and *Castor canadensis* and their ability to modify ecosystems', *Mammal Review* 35(3–4), 248–76 (2005), and *Rewilding and Flood Risk Management Briefing* (2016), available from Rewilding Britain: https://www.rewildingbritain.org.uk/assets/uploads/files/publications/Rewilding%20and%20Flood%20Risk%20Management%20briefing.pdf

230 In trial releases: Stringer, A. P. & Gaywood, M. J., 'The impacts of beavers *Castor* spp. on biodiversity and the ecological basis for their reintroduction to Scotland, UK', *Mammal Review* 46(4), 270–83 (2016).

230 a five-year study in Devon: Brazier, R. E., Elliott, M., Andison, E., Auster, R. E., Bridgewater, S., Burgess, P., Chant, J., Graham, H., Knott, E., Puttock, A. K., Sansum, P. & Vowles, A., *River Otter Beaver Trial: Science and Evidence Report* (2020), available from: https://www.exeter.ac.uk/creww/research/beavertrial/

230 a pair was released into Yorkshire . . . and a further release has followed in Exmoor, with another approved in Sussex:

https://www.forestryengland.uk/news/beavers-arrive-yorkshire-trial

https://www.theguardian.com/environment/2020/feb/01/beavers-uk-estate-owners-reintroduction-conservation-flooding

231 Released populations in Scotland and Devon draw tourists: Brazier, R. E., Elliott, M., Andison, E., Auster, R. E., Bridgewater, S.,

Burgess, P., Chant, J., Graham, H., Knott, E., Puttock, A. K., Sansum, P. & Vowles, A., *River Otter Beaver Trial: Science and Evidence Report* (2020), available from: https://www.exeter.ac.uk/creww/research/beavertrial/

231 One study estimated: Campbell, R. D., Dutton, A. & Hughes J., *Economic Impacts of the Beaver*, report for the Wild Britain Initiative (2007).

231 Landowners in Scotland: https://www.scotsman.com/news/environment/scots-farmers-rush-shoot-beavers-ahead-new-protection-law-1470879

232 The NGO Rewilding Britain suggests: *Rewilding and Climate Breakdown: How restoring nature can help decarbonise the UK* (2019). Report available from Rewilding Britain: https://www.rewildingbritain.org.uk/assets/uploads/Rewilding%20and%20Climate%20Breakdown%20-%20a%20report%20by%20Rewilding%20Britain.pdf

(We're done. Thanks for watching.)

ABOUT THE AUTHOR

Dr Tom Moorhouse is a conservation research scientist in Oxford University's Zoology department. His work has focused on water voles, and, more recently, the management of signal crayfish, hedgehog conservation, the impacts of wildlife tourism, and global demand for wildlife products.

Outside of conservation research, Tom is the author of award-winning children's fiction. He has also published articles based on his research, including the winner of the 2003 New Scientist New Millennial Science Writing Competition, *Reintroducing 'Ratty'*. He lives with his wife and daughter in Oxford and spends as much time as possible beside water.